Driving Emissions to Zero

Are the Benefits of California's Zero Emission Vehicle Program Worth the Costs?

Lloyd Dixon

Isaac Porche

Jonathan Kulick

RAND

Sponsored by
RAND Corporation and RAND Science and Technology

The research described in this report was conducted by RAND Science and Technology and sponsored jointly by the RAND Corporation and RAND Science and Technology.

ISBN: 0-8330-3212-7

Published 2002 by RAND
1700 Main Street, P.O. Box 2138, Santa Monica, CA 90407-2138
1200 South Hayes Street, Arlington, VA 22202-5050
201 North Craig Street, Suite 202, Pittsburgh, PA 15213-1516
RAND URL: http://www.rand.org/
To order RAND documents or to obtain additional information, contact Distribution Services: Telephone: (310) 451-7002; Fax: (310) 451-6915; Email: order@rand.org

PREFACE

The Zero Emission Vehicle (ZEV) program is a controversial part of California's strategy for meeting federal air quality standards. The program has been significantly modified multiple times since it was adopted by the California Air Resources Board in 1990 and is currently being challenged in court.

This report is an independent assessment of the costs and emission benefits of ZEVs and the other low-emission vehicles that manufacturers are allowed to use to meet ZEV program requirements. It reviews the program in the context of the overall strategy for reducing emissions in the greater Los Angeles area and makes recommendations for reform. The analysis and results presented should be of interest to government agencies, environmental groups, and automakers involved in developing policies to improve air quality in California.

This research is based on publicly available research reports, journal articles, newspaper stories, information available on the World Wide Web, and correspondence with experts in the field. It builds on other RAND work on air pollution issues in California. Related publications, all of which are available from RAND at nominal cost, include

- *California's Ozone-Reduction Strategy for Light-Duty Vehicles: Direct Costs, Direct Emission Effects, and Market Response,* Lloyd Dixon and Steven Garber, RAND, MR-695, 1996.
- *Fighting Air Pollution in Southern California by Scrapping Old Vehicles,* Lloyd Dixon and Steven Garber, RAND, MR-1256, 2001.
- *The Impact of Extending Emission Warranties on California's Vehicle-Repair Industry,* Lloyd Dixon, RAND, forthcoming, 2002.

This report was funded by RAND and RAND Science and Technology, a research unit within RAND. For more information about this report, contact:

Lloyd Dixon
RAND
1700 Main Street
Santa Monica, CA 90407-2138
TEL: 310.393.0411 x7480
FAX: 310.451.7062
Email: dixon@rand.org

CONTENTS

FIGURES

TABLES

SUMMARY

California has made significant progress in improving air quality in many parts of the state. However, substantial reductions in emissions of non-methane organic gases (NMOG) and oxides of nitrogen (NOx) are still needed to meet federal standards in California's South Coast Air Basin by 2010, as required by the Clean Air Act.[1] The South Coast Air Quality Management District (SCAQMD) and the California Air Resources Board (CARB) have adopted an aggressive strategy to reduce emissions. A controversial part of this strategy is the state's Zero Emission Vehicle (ZEV) program, which requires that auto manufacturers begin selling ZEVs starting in 2003. The ZEV program is a first step in achieving CARB's long-term goal of reducing emissions from California's motor vehicle fleet to zero. CARB believes that reliance on traditional gasoline-engine technology will not allow California to meet federal air quality standards.

This report examines whether ZEVs are a cost-effective way to achieve air quality standards in California. To this end, it examines the promise of technologies that could be used to satisfy ZEV program requirements. It examines the costs of ZEVs, the emission benefits that ZEVs generate, and the cost per ton of emissions reduced through the use of ZEVs. It reviews the ZEV program in the context of the state's overall strategy for reducing emissions in the South Coast Air Basin and compares the cost-effectiveness of ZEVs (as measured by cost per ton of emissions reduced) with that of other components of the overall strategy. It concludes with recommendations for policies on ZEVs and on California's strategy for controlling emissions from passenger cars and light-duty trucks more generally.

OZONE REDUCTION PLAN FOR THE SOUTH COAST AIR BASIN

To achieve federal ozone standards in the South Coast Air Basin, the SCAQMD estimates that NMOG and NOx emissions cannot exceed 413 and 530 tons per day, respectively. These levels are roughly one-half the emission rates observed in recent years. Regulations expected to achieve more than the required emission reductions for NOx have been adopted, but such is not the case for NMOG. Programs that reduce NMOG emissions by another 138 tons per day must still be found.[2]

---

[1]The South Coast Air Basin includes all of Orange County and the western, urbanized portions of Los Angeles, Riverside, and San Bernardino counties.
[2]Reductions in NOx can be substituted for reductions in NMOG to some extent.

The NMOG shortfall is very large, and it motivates CARB's vision of a zero emission fleet. This dire situation does not necessarily require a zero emission fleet however. Such an approach makes sense only if it costs less to eliminate the last bit of emissions from the on-road vehicle fleet than it does to reduce emissions from stationary and off-road sources. A fleet with very low emissions will undoubtedly be required to meet air quality standards in the South Coast Air Basin, but not necessarily a fleet with zero emissions.

To create a baseline with which to compare the costs of ZEVs, we gathered estimates of the cost-effectiveness of regulations that have recently been or are expected to be adopted in the South Coast. As shown in Table S.1, the cost per ton of NMOG plus NOx reduced—the cost-effectiveness ratio—varies greatly by emission source. To date, the most expensive regulations have been those imposed on gasoline engines for mobile sources—up to $33,000 per ton of NMOG plus NOx reduced. Costlier measures than those in Table S.1 may eventually be necessary to meet air quality standards in the South Coast. To achieve these standards at the least cost, however, policymakers should exhaust lower cost-per-ton alternatives before adopting expensive regulations on a particular source.

COSTS AND EMISSION REDUCTIONS OF TECHNOLOGIES FOR MEETING ZEV PROGRAM REQUIREMENTS

Manufacturers may use various types of vehicles to satisfy ZEV program requirements: ZEVs, partial zero emission vehicles (PZEVs), and gasoline hybrid electric vehicles (GHEVs). Large-volume auto manufacturers must meet a minimum portion of the requirements with ZEVs and have the option to satisfy other portions with PZEVs and GHEVs. PZEVs are gasoline internal combustion engine vehicles that must meet very stringent emission control standards; GHEVs are PZEVs that integrate an electric motor with the gasoline engine.

We calculated the costs and NMOG and NOx reductions of the various vehicle types relative to CARB's most stringent emission standards outside the ZEV program—i.e., the exhaust standard for a super ultra low emission vehicle (SULEV) and the near-zero evaporative emission standard. Incremental cost includes both the added cost of producing the vehicle and the difference in lifetime operating cost. We then combined the costs and emission reductions to yield estimates of the cost per ton of moving to progressively tighter emission standards.[3] The results are reported in Table S.2. The second column lists cost per ton in the near term, and the third lists cost per ton in high-volume production. The range given in the brackets reflects uncertainty in the underlying parameters.

---

[3]The vehicles rank as follows by emissions (in decreasing order): vehicles meeting SULEV exhaust and near-zero evaporative standards, PZEVs, GHEVs, and ZEVs.

Table S.1

Cost-Effectiveness Ratio of Recent and Expected Regulations to Reduce NMOG and NOx Emissions in the South Coast Air Basin ($ per ton of NMOG+NOx reduced)

Source Category	Dollars per Ton
Stationary Sources	
Consumer products	500-6,900
Other stationary sources	800-20,400
Mobile Sources	
On-road vehicles	
Gasoline engines	Up to 33,000
Diesel engines	400-800
Off-road vehicles	
Gasoline engines	Up to 20,600
Diesel engines	Up to 700

The high-volume estimates are based on designs that appear feasible given what is known about battery and fuel-cell technology. They also take into account the types of production processes that are feasible at high volume. Forecasting technological advancement is difficult and becomes more difficult the farther one looks into the future. We thus think it appropriate to interpret our high-volume estimates as the lowest level to which costs are expected to fall over the next 10 years or so given what is currently known about advanced vehicle technologies and manufacturing processes. Experience gained during manufacturing may reduce costs beyond our volume production predictions, but we do not expect these effects to be large over the next 10 years absent significant technological advances.

The lower costs in volume production can be thought of as the payoff society can expect after incurring high costs when a new technology is introduced. But in evaluating whether such an investment makes sense, society should consider not only the payoff, but also the costs incurred to achieve it. The final column of Table S.2 shows cost per ton when the higher cost of the vehicles produced in the near term is averaged in with the lower costs in volume production. The average is calculated based on assumptions about the number of vehicles produced in the near term, the decline in production cost over time (which is a function of production volume), and the number of vehicles produced once high-volume production is achieved. Our conclusions based on the results in Table S.2 follow.

Partial Zero Emission Vehicles

Our analysis suggests that PZEVs are an economical way to reduce NMOG and NOx emissions for passenger cars and smaller light-duty trucks.[4] We estimate that the cost of

---

[4]Light-duty trucks are trucks with loaded vehicle weight ≤3,750 pounds (CARB's LDT1 vehicle category).

Table S.2

**Cost-Effectiveness Ratio of Technologies That May Be Used to Satisfy ZEV Program
Requirements (ranges in brackets; $1000s per ton of NMOG+NOx reduced)**

Vehicle Type	In Near Term	High-Volume Production	Average Over Time
2003-2007			
Partial-zero emission vehicle (PZEV) relative to vehicle meeting SULEV exhaust and near-zero evaporative emission standards	[18, 71]	[18, 71]	[18, 71]
Gasoline hybrid electric vehicle (GHEV) relative to PZEV	[650, 1,800]	[-500, 90]	[-440, 180]
Full-function electric vehicle[a] (EV) relative to PZEV			
Nickel metal hydride batteries	[1,500, 2,300]	[260, 710]	[330, 810]
Lead acid batteries	[750, 1,500]	[12, 400]	[60, 470]
City EV[b] relative to PZEV			
Nickel metal hydride batteries	[1,400, 2,100]	[70, 370]	[150, 470]
Lead acid batteries	[850, 1,600]	[-80, 160]	[-20, 250]
2006-2010			
Direct hydrogen fuel-cell vehicle (DHFCV) relative to PZEV	[650, 1,300]	[-110, 270]	[-80, 310]

NOTE: Negative numbers indicate that incremental vehicle cost and thus cost per ton were negative.
[a]Full-function EVs are similar in size to many vehicles on the road today and are freeway capable.
[b]City EVs are much smaller than typical passenger cars that sell in the United States. They have limited top speed, acceleration, and range, and are not designed for regular freeway use.

reducing SULEV exhaust and near-zero evaporative emission levels to PZEV levels ranges from $18,000 to $71,000 per ton of NMOG plus NOx. The lower end of this range is less than the cost per ton of other regulations that have recently been adopted. The upper end of the range exceeds the costs of recent regulations, but it is plausible that cost per ton will need to rise this high before air quality standards are met.

Gasoline Hybrid Electric Vehicles

Whether GHEVs are a cost-effective way to reduce NMOG plus NOx emissions below PZEV levels remains uncertain. The outcome rests largely on whether the battery lasts the life of the vehicle. If the hybrid's maintenance costs are comparable to those of a PZEV, then GHEVs will be attractive. But if the battery needs replacing, cost may go as high as $180,000 per ton of emissions reduced.

Battery-Powered Electric Vehicles

Our analysis of BPEVs found that the cost of reducing emissions from PZEV levels to zero using EVs powered with nickel metal hydride (NiMH) batteries is high. NiMH batteries are most likely required for BPEVs with a 100-mile range.[5] We found that the cost per ton of emissions reduced was exceedingly high for full-function EVs using NiMH batteries between 2003 and 2007. Even in volume production, the cost per ton of NMOG plus NOx reduced would range from $260,000 to $710,000 per ton.

The situation is somewhat better for city EVs with NiMH batteries, but cost per ton still ranges from $70,000 to $370,000 in volume production. On a cost per ton basis, city EVs with NiMH batteries thus do not appear very attractive as a way to reduce emissions. What is more, city EVs are very different from vehicles on the road in California today. This raises questions about whether manufacturers will be able to sell substantial quantities at prices that cover their costs and whether the miles driven in city EVs will fully displace miles driven by conventional vehicles.

We also examined costs for full-function EVs and city EVs powered with lead acid (PbA) batteries. At the low end of our estimated range, which is based on cost and performance parameters that are plausible but optimistic, city EVs are attractive, and the cost per ton for full-function EVs is arguably not above what might be necessary to meet air quality standards. However, the case for EVs with PbA batteries is still not strong. First, the size and weight of these batteries make it difficult to produce vehicles with broad appeal. Second, the cost per ton may just as likely turn out to be very high. Our cost estimates top out at $470,000 and $250,000 per ton for full-function EVs and city EVs with PbA batteries, respectively, once the cost of vehicles produced before reaching high-volume production is included.

Our analysis leads us to conclude that battery-powered EVs do not appear to be an economical way to reduce NMOG and NOx emissions from PZEV levels. The main stumbling block is still battery cost and energy density, and after a decade of intense development, battery cost and performance remain well short of what is needed for a cost-effective vehicle. CARB's Battery Technical Advisory Panel concluded that major advances or breakthroughs that could reduce battery costs are unlikely through 2006 to 2008. Manufacturing costs could fall below our high-volume estimates as manufacturers gain production experience, but our analysis suggests that substantial reductions are possible only with material cost (battery design) breakthroughs.

Direct Hydrogen Fuel-Cell Vehicles

Direct hydrogen fuel-cell vehicles (DHFCVs) show much more promise than battery-powered EVs, but a great deal of uncertainty remains. Projections that fuel-cell system costs (excluding the fuel tank) can fall to $35 to $60 per kilowatt in volume production are well

_____
[5]A 100-mile range is also possible with lithium ion batteries, but this technology has shortcomings of its own.

grounded. Costs might be even lower if the designs with very low platinum loadings pan out, but it is too early to tell if this will be the case. Manufacturing experience may also reduce costs, although we conclude that these reductions may not be large, because most nonmaterial costs have been squeezed out of the low end of our predicted range by mature manufacturing techniques.

If fuel-cell costs do eventually fall to the lower end of the range ($35 per kilowatt), DHFCVs will be an attractive part of California's strategy for meeting ozone standards. However, there is little margin for error. A decline in costs to $60 per kilowatt would be a great technological achievement given where costs are now, but it would not be enough to reduce costs per ton to levels that might be required to meet air quality standards: Cost of reducing NMOG and NOx emissions from PZEV levels is $270,000 at the upper end of our predicted range in volume production.

Much has to be done to confirm the promise of DHFCVs. The performance and durability of designs with low precious metal loadings need to be tested in the real world. Given recent announcements by several automakers, it appears this testing will begin soon. The automated production processes assumed in volume production need to be validated. And importantly, a fueling infrastructure must be developed. Fuel-cell vehicles face a chicken-and-egg problem regarding fueling infrastructure: A sparse infrastructure limits the attractiveness of DHFCVs and the low number of DHFCVs limits the number of commercially viable fueling stations. We conclude from our analysis that it is too early to tell whether DHCFVs are an economical way to reduce emissions from PZEV levels to zero.

POLICY RECOMMENDATIONS

Our recommendations are directed at California policymakers and the CARB in particular. These recommendations are based on the overall social costs and benefits of the ZEV program. Some program costs might be spread outside California and thus may be of less concern to California policymakers. However, the interests of society as a whole are served by considering costs and benefits whether or not they occur in California. We chose to take this broader view.

Recommendation 1. Drop the Goal of Reducing Fleet Emissions to Zero

CARB should drop its goal of reducing emissions from the vehicle fleet to zero. These emissions do not have to be zero to meet federal air quality standards in California. Tremendous progress has been made in reducing emissions from internal combustion engines since the ZEV program was adopted in 1990, and ZEVs are not necessary for the foreseeable future. Rather than setting a goal of zero emissions for some sources, emission reduction targets should be based on the most cost-effective strategy for reducing emissions from all sources.

Recommendation 2. Require Passenger Cars and Smaller Light-Duty Trucks to Meet PZEV Emissions Standards

CARB should gradually require passenger cars and smaller light-duty trucks to meet PZEV emission standards. This might be done by gradually reducing the fleet-average NMOG exhaust requirement for these vehicles to PZEV levels and requiring them to meet the zero evaporative emission standard.[6] CARB should factor indirect emissions (those from fuel extraction, processing, and distribution) into the fleet-average NMOG requirement. This assures that vehicles such as gasoline hybrids, which may turn out to be an attractive way to reduce NMOG and NOx emissions, will be appropriately encouraged. CARB should also explore the costs of reducing the emissions of heavier light-duty trucks to PZEV levels.

Recommendation 3. Eliminate the ZEV Requirement

We think a strong case can be made for eliminating the requirement that manufacturers produce ZEVs. We found that ZEVs do not look attractive in terms of cost-effectiveness, but we also looked at some other factors, ones that cannot be easily quantified. The most important of these include

- technology development
- insurance against disappointment in emission control technology for gasoline vehicles
- reductions in carbon dioxide emissions and dependence on foreign oil
- lack of alternatives for reducing emissions.

Many potential costs of the ZEV requirement are also difficult to quantify. Below we review our evaluations of these factors and conclude with what we see as the implications for ZEV policy.

Technology Development. One reason to keep the ZEV requirement is that it may force technology development that would not otherwise occur. Fuel-cell programs have developed substantial momentum, but the fact is that we simply do not know what would happen to fuel-cell research and development if the ZEV requirement were scrapped. On the other hand, keeping the ZEV requirement may push the wrong technology. Battery-powered EVs have turned out to be a bad choice, and DHFCVs may turn out the same. For example, hydrogen infrastructure problems might prove so significant that fuel-cell vehicles with on-board reformers that generate hydrogen from gasoline or methanol (along with some emissions) might make sense. It may turn out that

---

[6]CARB requires that new vehicles sold by each manufacturer meet a fleet-average NMOG requirement. Each different exhaust emission category (the SULEV exhaust standard is one such category) is assigned an NMOG factor, and the average across all vehicles sold by a manufacturer during a year is its fleet-average NMOG.

vehicles that run on compressed natural gas or further refinements in gasoline internal combustion engine vehicles will prove to be more cost-effective ways to reduce emissions.

Insurance Against Disappointments in Emission Control Technology for Gasoline Vehicles. The ZEV requirement provides some insurance against the possibility that the lifetime emission profiles of PZEVs and other very clean gasoline vehicles do not turn out to be as low as currently projected. PZEVs have yet to be sold, and it will be many years before their in-use emissions can be verified. It may be the case that the on-board diagnostics systems that monitor emission performance do not work well when emissions are at such low levels, or that the emission control system deteriorates faster than expected. The important question here, however, is not whether the ZEV program provides some insurance, but at what cost. Other types of insurance may be more cost-effective—for example, research on how to further reduce emissions from stationary sources and from off-road and diesel vehicles.

Reductions in Carbon Dioxide Emissions and Dependence on Foreign Oil. Even if ZEVs do not look like a cost-effective way to reduce NMOG and NOx emissions, the ZEV requirement pushes the development of technologies that can reduce carbon dioxide emissions and dependence on foreign oil. The value of the ZEV requirement for achieving these two reductions is uncertain, however, in that there may well be much more cost-effective ways to reach the same ends. For example, to reduce carbon dioxide emissions, it may be more cost-effective to raise the Corporate Average Fuel Economy (CAFE) standard than to require ZEVs. To reduce dependence on foreign oil, it may make more sense to switch to alternative-fuel vehicles that produce some emissions, such as vehicles that run on compressed natural gas.

Lack of Alternatives for Reducing Emissions. ZEVs may still make sense as part of California's ozone reduction strategy if the alternatives for reducing emissions look even less promising. We have not developed a detailed plan that would reduce NMOG and NOx emissions to the levels needed to meet ozone standards in the South Coast Air Basin. However, we note that a number of alternatives to ZEVs appear to be available. First, our analysis suggests that PZEVs are cost-effective, and moving the fleet to PZEVs would substantially reduce light-duty vehicle emissions. Second, our analysis shows that substantial emissions are still being produced by sources other than light-duty vehicles—particularly, sizable NMOG emissions from solvent use in the South Coast Air Basin. Policymakers should explore possibilities for reducing emissions from these sources before requiring ZEVs.

Uncounted Potential Costs. Our estimates of cost-effectiveness do not include some potential costs. First, the program may cause the prices of new vehicles to increase and thus slow the sales of new vehicles, thereby causing the average age and emissions of the vehicle fleet to increase. While this feedback is possible in principle, we found that there is a great deal of uncertainty about its size. Second, reductions in operating and maintenance costs may cause drivers to travel more, offsetting the emission benefits of some vehicles. Given the ZEV's range and fueling infrastructure limitations, however, it seems unlikely that ZEVs will encourage more travel. Finally, interactions with CARB's fleet-average NMOG requirement and the federal CAFE standards will likely offset some of the ZEV emission reductions. ZEVs are included in the calculation of fleet-average NMOG emissions and fuel economy and thus allow the non-ZEV portion of the fleet to have higher emissions or lower fuel economy than it would otherwise. In principle, these regulations can be rewritten, but there are often political obstacles to doing so. While it is difficult to assess the overall importance of these uncounted costs, we know they work against any uncounted potential benefits.

Synthesis. There is room for disagreement over whether the difficult-to-quantify benefits outweigh the difficult-to-quantify costs, but we conclude, overall, that the ZEV program should be ended. What concerns us most about the ZEV requirement is that it focuses on a very narrow set of technologies in its aim to reduce air pollution in California. The focus on zero emission technologies seems particularly inappropriate given that ZEVs are not needed to meet air quality standards and that lower-cost alternatives for reducing emissions appear to be available. We also do not find the arguments that ZEVs can reduce carbon dioxide emissions or dependence on foreign oil to be convincing. These two social goals may well be important, but much more needs to be done to show that ZEVs are a cost-effective way to achieve them.

Recommendation 4. If the ZEV Requirement Cannot Be Eliminated, Delay It or Reduce the Number of Fuel-Cell Vehicles Needed to Satisfy It in the Early Years

As it stands now, the ZEV program will force at least some manufacturers to produce BPEVs during its initial years. BPEVs are expensive to produce and show little long-term promise, so it makes no sense to continue investing in this technology. By delaying the program, CARB would allow time to evaluate the promise of DHFCVs, the only other zero-emission technology that appears viable for the foreseeable future.

If CARB does not want to delay the program because it believes doing so will substantially reduce incentives for bringing the potentially promising direct hydrogen fuel-cell technology to the market, it might, as an alternative, significantly reduce the number of DHFCVs needed to satisfy the ZEV requirement. This could be done by increasing the number of ZEV credits generated by each DHFCV. A substantial increase would prevent a large number of battery-powered or fuel-cell vehicles from being put on the road, but would still allow CARB and the manufacturers to better understand the real-world potential of fuel-cell vehicles.

Whether the program is delayed or the required number of DHFCVs is reduced by increasing the credit multiplier, CARB should periodically assess the progress being made in

DHFCVs. If it becomes clear that DHFCVs are not a cost-effective way to reduce emissions, the ZEV program should be abandoned. If DHFCVs live up to their potential, the credit multiplier could be gradually reduced and production increased.

Recommendation 5. Focus on Performance Requirements, Not Technology Mandates

CARB should focus on setting performance requirements and let the automakers determine how best to achieve them. It should continue to set very stringent fleet-average emission requirements for the new vehicle fleet, but there is no need to set allowable emissions to zero or to require manufacturers to meet the average in part with ZEVs.

California has made remarkable progress in improving its air quality over the last 30 years. However, much remains to be done. Policymakers should seek to facilitate and encourage technological transformation while allowing the flexibility needed for different technologies to compete. Allowing such flexibility would be a departure from the ZEV program, but one well worth pursuing.

ACKNOWLEDGMENTS

Many people made valuable contributions to this report. Technical reviews of the draft report were provided by Frank Camm at RAND, Howard Gruenspecht at Resources for the Future, and Tim Lipman at the Renewable and Appropriate Energy Lab, UC Berkeley. We thank them for their detailed, insightful comments, which led to the report being improved considerably.

Steve Albu, Paul Hughes, Mike McCarthy, and Chuck Shulock at the California Air Resources Board (CARB) provided very thoughtful feedback on the draft report. Cynthia Marvin and Andrew Panson at CARB and Laki Tisopulos at the South Coast Air Quality Management District provided documentation and interpretation of recent amendments to the State Implementation Plan for Ozone. Alec Brooks and Tom Gage at AC Propulsion, David Friedman at Natural Resources Defense Council, Karl-Heinz Hauer at Xcellvision (in Wolfsburg, Germany), and Ben Knight at Honda helped us find and interpret information on battery and fuel-cell technologies.

Within RAND, Jim Bartis helped us develop estimates of electricity and hydrogen costs and provided useful comments on an interim draft. Paulette Middleton explored issues that underlie the vehicle emissions estimates, and Mark Bernstein helped frame the issues. Steve Rattien, director of RAND's Science and Technology Unit, provided helpful feedback during the project. Debra Knopman managed the review process and made valuable suggestions for improving the report. The document was edited by Jeri O'Donnell and corrected and formatted by Pat Williams and Joanna Nelsen. We thank them all for their contributions.

ABBREVIATIONS

Abbreviation	Definition
ADL	Arthur D. Little, Inc.
Ah	Ampere-hour
ANL	Argonne National Laboratory
ATPZEV	advanced technology partial zero emission vehicle
BPEV	battery-powered electric vehicle
BTAP	Battery Technology Advisory Panel
BTU	British thermal unit
CAFE	Corporate Average Fuel Economy
CALCARS	California Light Duty Vehicle Conventional and Alternative Fuel Response Simulator
CARB	California Air Resources Board
cc	cubic centimeter
cm	centimeter
CMPEG	California miles per equivalent gallon
DFMA	design for manufacturing and assembly
DHFCV	direct hydrogen fuel-cell vehicle
DOE	Department of Energy
DTI	Directed Technologies, Inc.
EV	electric vehicle
FCTAP	Fuel Cell Technical Advisory Panel
FMVSS	Federal Motor Vehicle Safety Standards
GHEV	gasoline hybrid electric vehicle
gm	gram
hp	horsepower
I&M	inspection and maintenance
ICE	internal combustion engine
ICEV	internal combustion engine vehicle
kg	kilogram
kW	kilowatt; unit of power equal to 1.34 horsepower
kWh	kilowatt-hour; a measure of energy storage
LDT	light-duty truck

LDT1	light-duty truck with loaded vehicle weight of 3,750 pounds or less
LDT2	light-duty truck, heavier than LDT1 but less than 8,500 pounds gross vehicle weight
LDV	light-duty vehicle
lpg	liquid petroleum gas
MEA	membrane electrode assembly
MMBTU	million British thermal units
MOA	memorandum of agreement
MSRP	manufacturer's suggested retail price
NiCd	nickel cadmium
NiMH	nickel metal hydride
NMOG	non-methane organic gases
NOx	oxides of nitrogen
O&M	operations and maintenance
OBD	on-board diagnostics system
OEM	original equipment manufacturer
PbA	lead acid
PEM	proton exchange membrane
PM10	particulate matter with diameter less than 10 microns
psi	pounds per square inch
PZEV	partial zero emission vehicle
R&D	research and development
SCAQMD	South Coast Air Quality Management District
SIP	state implementation plan
SULEV	super ultra low emission vehicle
SUV	sport utility vehicle
tpd	tons per day
ULEV	ultra low emission vehicle
ZEV	zero emission vehicle

1. CALIFORNIA'S AIR POLLUTION PROBLEM AND THE ZEV PROGRAM

California has made significant progress in improving air quality in many parts of the state. The improvement has been particularly dramatic in the South Coast Air Basin,[1] where the number of days that ozone levels exceeded federal standards fell from over 150 a year in the 1980s to 41 in 1999 (SCAQMD, 2001). The improvement in the South Coast has been due to increasingly stringent regulations being imposed on a wide range of stationary and mobile emission sources. However, even though much progress has been made, emissions of non-methane organic gases (NMOG) and oxides of nitrogen (NOx) must still be substantially reduced to meet federal standards by 2010, as required by the Clean Air Act.

The South Coast Air Quality Management District (SCAQMD) and the California Air Resources Board (CARB) have adopted an aggressive strategy to reduce emissions. A controversial part of this strategy is the state's Zero Emission Vehicle (ZEV) program. Starting in 2003, manufacturers that sell more than 60,000 light-duty vehicles (LDVs)—i.e., passenger cars and light-duty trucks (LDTs)—in the state must begin selling ZEVs. The number of ZEVs required is modest at first and then ramps up gradually over time. The ZEV program is a first step in achieving CARB's long-term goal of reducing vehicle emissions to zero. CARB believes that reliance on traditional gasoline-engine technology will not allow California to meet air quality requirements, and in its vision of the future, "the entire vehicle fleet will produce zero tailpipe emissions and will use fuels with minimal 'fuel cycle' emissions" (CARB, 2000b, p. 2).[2]

This report examines whether the ZEV program and the goal of a zero emission fleet more generally make sense for California. To this end, it examines the promise of the technologies that can be used to satisfy ZEV program requirements. It looks at the costs of ZEVs and the emission benefits they generate. It also reviews the ZEV program in the context of the overall strategy for reducing emissions in the South Coast Air Basin, and it compares the cost-effectiveness of ZEVs with that of other components of the overall strategy.

The rest of this section describes the ZEV program, reviews the main arguments for and against ZEVs, discusses the contributions of this study to the debate, and provides an outline of the report.

---

[1]The South Coast Air Basin includes all of Orange County and the western, urbanized portions of Los Angeles, Riverside, and San Bernardino counties.

[2]Fuel-cycle emissions occur during vehicle refueling and the production and transportation of fuel.

1.1 ZEV PROGRAM

CARB first adopted the ZEV program in 1990. It required that starting in 1998, 2 percent of passenger cars and smaller LDTs sold by large-volume manufacturers in California had to be ZEVs.[3] The requirement then rose to 5 percent in 2001 and to 10 percent in 2003. ZEVs were defined as vehicles that produce zero exhaust emissions under all operating conditions. As we discuss in subsequent sections, battery-powered electric vehicles (BPEVs) and direct hydrogen fuel-cell vehicles (DHFCVs, which are fueled with hydrogen gas) are the only ZEVs considered to be technically feasible for commercial production.

In 1996, after a series of public hearings and intense public debate, CARB concluded that ZEVs would not be commercially viable by 1998 and replaced the ZEV requirements between 1998 and 2003 with a memorandum of agreement (MOA) with each large-volume auto manufacturer. Under the MOAs, manufacturers were required to place more than 1,800 advanced BPEVs in California between 1998 and 2000. The 10 percent ZEV requirement for 2003 and beyond remained in place.

In 1998, CARB revised the program a second time, allowing manufacturers to satisfy at least part of the requirements with extremely clean, but not zero emission, vehicles called partial zero emission vehicles (PZEVs). PZEVs generate ZEV credits that can be used to reduce the number of ZEVs that a manufacturer must produce. Five conventional gasoline-powered PZEVs could substitute for one ZEV.[4] Large-volume manufacturers could satisfy up to 60 percent of program requirements with PZEVs.

The most recent revision of the program occurred in January 2001 (CARB, 2001a). The 2001 modifications

- Create a vehicle category called *advanced technology partial zero emission vehicles* (ATPZEVs) that manufacturers can use to fulfill an additional 20 percent of the requirement above the 60 percent that can be met with PZEVs. ZEVs can now fulfill as little as 20 percent of the requirement.
- Increase the credit multiplier based on the range a ZEV travels on a single charge. The number of credits now rises from one to 10 as the range increases from 50 to 275 miles.

---

[3]In 1990, CARB defined large-volume manufacturers as those that sold more than 35,000 light- and medium-duty vehicles per year (CARB, 2001e, p. 4). There were seven such manufacturers at that time.

[4]Manufacturers were required to produce ZEV credits equal to the ZEV program percentage requirement (i.e., 10 percent) times their California sales of passenger cars and smaller LDTs. PZEVs generated 0.2 credits, and certain types of ZEVs (e.g., ZEVs with a relatively large range on a single charge) could generate multiple credits.

- Phase in the maximum proportion of the program requirements that can be met with PZEVs.
- Gradually increase the ZEV program percentage from 10 percent between 2003 and 2008 to 16 percent in 2018 and subsequent years.
- Starting in 2007, include heavier sport utility vehicles, pickups, and vans in the base used to calculate program requirements.[5]

These modifications substantially reduce the number of ZEVs required during the first years of the program, but they ramp up the number of vehicles required in subsequent years.

Only large-volume manufacturers (now defined as those with passenger car plus LDT1 sales exceeding 60,000[6]) are subject to the ZEV and ATPZEV portions of the program. There are currently six large-volume manufacturers (the so-called Big 6): DaimlerChrysler, Ford, General Motors, Honda, Nissan, and Toyota. Intermediate-volume manufacturers can meet the entire program requirement with PZEVs, and small-volume manufacturers and independent low-volume manufacturers (those with annual sales of 4,500 or fewer and between 4,501 and 10,000, respectively) are not subject to the program.

States have the right to choose between the U.S. Environmental Protection Agency's (EPA's) or California's LDV regulations. Four northeastern states (Maine, Massachusetts, New York, and Vermont) adopted California's regulations but are considering or have adopted modifications. In January 2002, New York delayed its 2 percent ZEV requirement from 2003 to 2005 (Perez-Peña, 2002), and Massachusetts delayed its ZEV requirement until 2007 (Daniel, 2002). Vermont's governor has sought to scrap his state's ZEV requirements but has been rebuffed by the state legislature; Maine has already scrapped its requirements (Mace, 2002).

1.2 STATE OF THE DEBATE ON ZEV PROGRAM

The ZEV program has been hotly contested since its adoption in 1990. Arguing for the program have been environmental groups such as the Natural Resources Defense Council, the Union of Concerned Scientists, and the Sierra Club, as well as electric utilities such as Southern California Edison. CARB staff have also argued strongly in favor of the program, although some have worried that it will detract from their decades of success in controlling emissions from

---

[5]Prior to 2007, the program applies only to sales of passenger cars and smaller LDTs—i.e., LDTs with a loaded vehicle weight of 3,750 pounds or less (LDT1s). Most LDTs have a loaded vehicle weight greater than 3,750 pounds and a gross vehicle weight less than 8,500 pounds (the cutoff of the LDT category). These heavier LDTs are called LDT2s (see CARB, 2001e, p. 30).

[6]CARB, 2000c, p. 21.

gasoline engines. Arguing against the program have been the U.S. and Japanese auto companies and oil companies.[7]

The contesting parties agree on some issues. Most acknowledge that the cost per ton of emissions reduced by the program will be very high during its first years, for two reasons. First, the cost of BPEVs is currently high, and their limited vehicle range means that consumers will most likely not be willing to pay much of a premium (if any) for them. Vehicles using direct hydrogen fuel cells may also be able to meet the program requirements, but they are likely to be available only in pilot-program quantities before 2006, and their initial costs will likely be high as well. Second, the program's early years will yield only small emission reductions because of the limited number of ZEVs required relative to overall vehicle sales and because the emissions of the gasoline vehicles they will displace are very low.

Agreement ends here, however. Automakers assert that the ZEV program will actually increase emissions because it will cause new vehicle prices to rise, which will then slow both the sales of new, cleaner vehicles and the retirement of older, higher-polluting vehicles (Harrison et al., 2001); CARB strongly contests this argument (CARB, 2001d). There is also disagreement on long-term ZEV costs. CARB concludes that "under an optimistic but nonetheless plausible scenario, battery EVs could become cost-competitive with conventional vehicles on a lifecycle cost basis. This scenario assumes [annual] volume production of more than 100,000 ZEVs" (CARB, 2000b, p. vii). Automakers argue that even at high-production volumes, BPEVs will remain more expensive than conventional vehicles, and that even though fuel-cell technology is promising, many issues have yet to be resolved.

Two fundamental beliefs underlie CARB's continued support of the program in spite of high initial costs and limited evidence that ZEVs will be cost-effective in the long run. A review of these beliefs illustrates the issues that need to be examined in a review of the ZEV program.

First, CARB believes that "zero-emission technology is necessary to achieve the state's public health goals" (CARB, 2000b, p. i). As CARB sees it, emissions must be reduced to zero where feasible to balance emissions from other sources that will continue to increase with relentless population and economic growth. The goal is a zero emission fleet, and the ZEV program is the first step in the quest.

_____

[7]According to CARB staff: "Estimates presented by vehicle manufacturers, based on cost and emission benefit information from the August 7 Biennial Review staff report, indicated that at least in the early years of the program the dollars spent per ton of pollutant reduced under the ZEV program will be much higher than for any other ARB regulatory measure. Despite this information, which was not disputed by staff, the Board voted unanimously to maintain the program" (CARB 2000c, p. 41).

Second, CARB believes that the program will induce the development of zero emission technologies.[8] CARB has been requiring tighter emission standards on internal combustion engines for decades, and significant technological advances have dramatically reduced emissions and the cost of emission controls. The logic is that the zero emission requirements should be no different: Once manufacturers put their minds to it, they will discover the breakthroughs necessary to make electric vehicles (EVs) work. The breakthroughs may come for either BPEVs or DHFCVs. CARB has argued that even if technological breakthroughs do not materialize for batteries, DHFCVs will someday be available at attractive costs.

As a consequence of these two fundamental beliefs, CARB is not much concerned that the ZEV program has high costs per ton of emissions reduced in the near term. Society must invest now in order to see technological improvements later. What is more, the ZEV program is a "transformative leap forward," and "[g]iven the sweeping nature of ZEVs' environmental, energy and societal effects, it is reasonable to expect that the program will be more expensive in its early years than more limited measures" (CARB, 2000b, p. ix). In other words, a very large investment is warranted because the payoff will be very large. It is telling that nowhere does CARB explicitly calculate the cost-effectiveness of the ZEV program in the near term.

Nor are the modest emission reductions that CARB predicts for the ZEV program much cause for concern. These reductions are based on scenarios in which ZEVs account for a small proportion of new car sales. Once ZEV technology matures, the proportion will increase and so will emission reductions. CARB forcefully contests automaker claims that the ZEV program will cause fleetwide (both new and old vehicle) emissions to increase in the short run, and argues that even if they were to increase, they would decline eventually as older gasoline vehicles exit the fleet.

CARB also raises several secondary issues in its defense of the program. It argues that ZEVs will increase the diversity of California's transportation energy system, reducing the state's dependence on foreign oil and contributing to a more stable transportation fuels market. CARB also notes that even though its principal regulatory responsibilities are to reduce the levels of ozone and other criteria air pollutants, ZEVs will reduce greenhouse gas emissions as well. Finally, CARB believes that ZEVs can help California's economy by increasing the demand for services offered by the many advanced technology firms in the state (CARB, 2000b, p. viii).

---

[8]"The tremendous progress [in EV technology] that has been seen [since the program was adopted in 1990] can at least in part be attributed to the existence of the ZEV requirement, and staff believes that maintaining this requirement will accelerate the pace at which true zero technologies are commercialized" (CARB, 2000c, p. 8).

1.3 THIS STUDY'S CONTRIBUTION

This report examines several of the key beliefs and assumptions underlying arguments for and against the ZEV program. First, we explore the program's role in the overall strategy for reducing emissions of ozone precursors in the South Coast Air Basin. We examine whether a zero emission LDV fleet is indeed necessary to achieve air quality standards, and we review available information on the costs of reducing emissions from different sources. This inquiry allows us to better understand the alternatives to reducing emissions in the LDV to zero.

Second, we integrate findings from previous studies to evaluate the costs, emission benefits, and cost-effectiveness of the technologies that may be used to satisfy the ZEV program. This synthesis illuminates the attractiveness of using such technologies to meet air quality standards.

Third, we examine several of the program's indirect (or ancillary) benefits. In particular, we examine arguments that the ZEV program has induced the development of zero emission and advanced vehicle technology more generally and that it will diversify California's transportation energy base. Such benefits are not easy to quantify and are not included in typical estimates of the program's cost-effectiveness. The difficulty of quantifying any such benefits, however, should not dictate their exclusion from an evaluation of the program.

This study's final contribution is that it integrates findings on the key assumptions and beliefs underlying the arguments for and against the ZEV program and assesses whether the program makes sense for California.

1.4 RESEARCH APPROACH

Our approach in this study was to investigate the promise of the advanced technologies that automakers are pursuing to meet the ZEV program requirements. We did not attempt to make detailed predictions of the overall costs and emission reductions that will occur as automakers comply with the program. Such predictions are difficult because of the program's complexity and its interactions with other regulations on LDV emissions.[9] We examined what society can expect to gain from the availability of these advanced technologies, rather than whether their benefits will be diminished or amplified by other elements of the system.

---

[9]For example, manufacturers must meet a fleet-average NMOG requirement for the new vehicles they sell in a given year. ZEVs are included in the calculation of the NMOG average, so emissions from other new vehicles can be higher than they would otherwise be, and emission reductions from the ZEV program are thus smaller than a direct comparison of the emissions of ZEVs and other vehicles would suggest.

This research is based on publicly available research reports, journal articles, newspaper stories, information available on the World Wide Web, and correspondence with experts in the field.

As is shown in subsequent sections, the costs of ZEVs depend importantly on production volumes. We thus first predict costs during the first five years of the program (2003 through 2007) using production volumes that may plausibly occur over this period. The production volumes reflect various strategies that manufacturers may use to meet program requirements. We do not construct an exhaustive list of scenarios; rather, we construct scenarios that reflect plausible production volumes for different types of vehicles. We then predict the costs at high-volume production given what is known about ZEV technologies and manufacturing processes. These predictions are meant to give an indication of what society can expect of these technologies after a high-cost introductory period.

We combine cost predictions with estimates of the emission benefits of the various technologies to calculate the cost per ton of emissions reduced. This measure of cost-effectiveness, referred to as the *cost-effectiveness ratio*, is used to evaluate the attractiveness of ZEVs relative to other approaches for reducing emissions. To achieve air quality standards at the least cost, policymakers should exhaust lower cost-per-ton alternatives before adopting expensive regulations on a particular source.

Our analysis is concerned with the *overall* social costs and benefits of the requirements for advanced vehicle technologies; we are not concerned with the distribution of costs and benefits among auto producers, consumers, and the government. We also do not address the distribution of costs and benefits inside and outside California. Past studies have shown how the costs of earlier versions of the program were spread among various stakeholders and inside and outside California (e.g., Dixon and Garber, 1996). Our analysis looks at the costs to society as a whole (worldwide costs) of reducing emissions in California. These may not be the costs that California policymakers consider when making decisions, but decisions based only on costs to Californians may overlook important consequences. Because we examine overall social costs, we exclude gasoline and electricity taxes in our calculations of vehicle operating costs, since these are transfer payments between consumers and government. Similarly, we ignore government subsidies of electric and other advanced technology vehicles and utility subsidies of electricity rates for EV charging.

1.5 REPORT ORGANIZATION

To illustrate the role of the ZEV program in California's clean air strategy, we examine in Section 2 the emission reduction targets for the various source categories and the success so far in adopting regulations to achieve these targets. We also review the cost-effectiveness of pollution control measures that have recently been adopted or are expected to be adopted in coming years. Such a review provides a benchmark for comparing the cost-effectiveness of ZEVs.

Section 3 provides an overview of the ZEV, PZEV, and ATPZEV technologies that automakers may use to satisfy the program's requirements. It also describes the production volume scenarios we used in our analysis.

Section 4 examines the costs of ZEVs, PZEVs, and ATPZEVs. It predicts costs in the early years of the program and in volume production. It also estimates the costs of producing the vehicles, as well as the costs of operating and maintaining them over their lifetimes.

Section 5 develops estimates of the emissions of ZEVs, PZEVs, and ATPZEVs relative to the cleanest vehicles that CARB will require outside the ZEV program. This analysis allows us to determine what the additional costs of ZEVs, PZEVs, and ATPZEVs buy in terms of emission reductions.

In Section 6, we combine Section 4's cost estimates with Section 5's emission estimates to calculate the costs of ZEVs, PZEVs, and ATPZEVs per ton of emissions reduced. Since these cost-effectiveness estimates do not reflect all considerations that should enter an evaluation of ZEVs, we also examine other potential costs and benefits.

Section 7 summarizes our key results and draws conclusions about the attractiveness of the ZEV program for reducing emissions of ozone precursors in California.

Five appendices detail several of our calculations and provide support for assumptions used in our analysis.

2. OZONE REDUCTION PLAN FOR THE SOUTH COAST AIR BASIN

The ZEV program is one element of an aggressive plan to reduce ozone levels in the South Coast Air Basin. This plan involves substantial emission reductions from both stationary and mobile sources. In this section, we first review the emission targets for each source category to show how the targets for mobile sources compare to those for stationary sources and to put CARB's goal of a zero emission fleet in perspective. Second, we examine the strategy in place for achieving the targeted emission reductions so as to determine how complete the strategy is for meeting the emission targets and the urgency with which additional ways of reducing emissions must be found. Finally, we present estimates of the cost-effectiveness of emission reduction measures that have recently been adopted or are expected to be adopted in the coming years. These estimates allow us to compare the cost-effectiveness of ZEV and partial zero emission vehicle (PZEV) technologies with the cost-effectiveness of reducing emissions from other sources.

2.1 REGULATORY RESPONSIBILITY FOR REDUCING EMISSIONS IN THE SOUTH COAST AIR BASIN

Several levels of government are involved in crafting and implementing the plan to reduce ozone levels in the South Coast Air Basin. The South Coast Air Quality Management District (SCAQMD) has responsibility for controlling stationary point sources.[1] The South Coast Association of Governments promulgates "transportation control measures" that attempt to reduce vehicle miles traveled by shifting people from single-occupant vehicles to other modes of transportation. CARB has responsibility for most on- and off-road mobile sources, but the U.S. EPA retains authority for trains, ships, airplanes, and certain categories of off-road equipment (e.g., tractors with more than 175 horsepower) and sets regulations for passenger cars and trucks sold outside California. These federal regulations affect emissions in the South Coast because some of these vehicles are driven in or migrate to Southern California.

---

[1]Emission sources are divided into two major categories: stationary and mobile. Stationary sources are in turn divided into two main subcategories: point and area sources. "Point sources are generally large emitters with one or more emission sources at a permitted facility with an identified location (e.g., power plants, refinery boilers). Area sources generally consist of many small emission sources (e.g., residential water heaters, architectural coatings) which are distributed across the region" (SCAQMD, 1996, p. 3-2). Mobile sources are divided into two subcategories: on-road and off-road sources. Passenger cars and heavy-duty trucks are examples of on-road sources. Trains, ships, airplanes, and off-road vehicles such as dirt-bikes, tractors, and construction equipment are examples of off-road sources.

The SCAQMD takes the lead in assembling the air quality management plan for the South Coast Air Basin. This plan projects emission levels over time given the regulatory controls currently in place and also specifies target emission levels for each source. Presumably, these emission targets are the result of ongoing negotiations and jockeying among different levels of government and stakeholder groups. The plan uses airshed models to demonstrate that the target emission levels will achieve air quality standards in the South Coast. The emission inventory and target emission levels were last updated in 1996 and released as the *1997 Air Quality Management Plan*, which we refer to as the 1997 Plan (SCAQMD, 1996).

The air quality management plan for the South Coast describes the control measures that will be adopted to achieve the target emission levels. The SCAQMD last updated the strategy for reducing emissions from stationary point sources in its 1999 amendments to the 1997 Plan, which we refer to as the 1999 Plan (SCAQMD, 1999). The basic structure of CARB's strategy to reduce emissions from mobile sources and consumer products is detailed in its 1994 state implementation plan (SIP) for ozone (CARB, 1994); updates to this strategy are described in CARB, 2000d.

2.2 TARGET EMISSION LEVELS IN THE SOUTH COAST

Dramatic emission reductions will be required to achieve federal ozone standards in 2010.[2] To meet the standards, the 1997 Plan estimates that emissions of non-methane organic gas (NMOG) cannot exceed 413 tons per day and emission of oxides of nitrogen (NOx) cannot exceed 530 tons per day during the summer months (SCAQMD, 1996, pp. 4-32, 5-23).[3] These levels are far below recent levels. Given regulations adopted through September 1996, the SCAQMD projected that NMOG emissions averaged 937 tons per day and that NOx emissions averaged 916 tons per day during summer 2000 (SCAQMD, 1996, p. 3-15). Thus, NMOG and NOx emissions must be reduced by roughly 50 percent from recent levels to meet the ozone standards.

---

[2]The EPA issued a stricter ozone standard in 1997, which was challenged in court. Recently, the U.S. Supreme Court confirmed EPA's authority to issue the new standards but found its implementation policy for the new standard unlawful (U.S. Supreme Court, 2002). Thus, a number of implementation issues related to the new standard have yet to be resolved. Greater emission reductions in the South Coast will be required to achieve the new standards, if and when implementation issues are resolved. The California state standards are stricter than the pre-1997 federal standard and similar to the new federal standard. Most attention has been on the federal standards. California has not yet vigorously pursued enforcement of the state standards.

[3]The ratio of NOx and NMOG emissions needed to meet air quality standards is somewhat flexible. Thus, the target for NMOG emissions relative to NOx emissions is also a policy decision. The extent to which the costs of reducing NMOG relative to NOx entered the determination of the emission targets deserves further investigation.

The 1997 Plan calls for substantial emission reductions across all types of sources. To quantify the reductions required from additional regulations, it compares emissions in 2010 given all regulations adopted through September 1996 (baseline emissions) with emission targets in 2010 for each source (controlled emissions). This comparison is made because emissions will fall between now and 2010 due to implementation of regulations that have already been adopted. Tables 2.1 and 2.2 report the baseline emissions and emission reduction targets for NMOG and NOx, respectively. Emissions are reported in tons per day (tpd) for weather conditions equivalent to those on the first day of the August 1987 smog episode.[4] The 2010 controlled, or target, emissions thus do not exactly match those for the summer inventory.

In percentage terms, the target NMOG emission reductions are quite similar across the three major categories (stationary, on-road mobile, and off-road mobile), varying between 50 and 57 percent. Emissions from stationary sources accordingly account for about two-thirds of total NMOG emissions both before and after the additional reductions. Within stationary and off-road mobile sources, however, there is considerable variation in the target emission reductions.

There is more variation in the target percentage reductions for NOx, with off-road mobile shouldering the largest reductions (39 percent). The higher percentage reduction for off-road mobile sources reflects the less stringent regulation of these sources in the past. In contrast to NMOG, roughly 85 percent of the NOx emissions in 2010 are planned to be from mobile sources—the same both before and after the additional reductions.

The numbers in these tables prompt two observations. First, CARB's goal is a zero emission (presumably on-road) mobile fleet. But a zero emission fleet is obviously not necessary to meet NMOG and NOx emission targets. There is a set quantity of emissions, determined by the weather and geography in the South Coast, that can be allocated to the emission sources. Whether it makes sense for mobile emissions to be zero depends on the relative costs of reducing emissions from the various sources.

Second, the emission reductions in the 1997 Plan are not necessarily the least costly way to reduce emissions. Presumably, the 1997 Plan reflects to some extent the relative difficulty of reducing emissions from different sources. However, incomplete information on the costs and emission effects of prospective regulations makes it difficult to construct an optimal plan. Moreover, the political compromise usually needed to settle on a plan can yield results that are far from optimal. The similar proportionate cutbacks for stationary and mobile sources may be the result of political compromise rather than being the least costly way of reducing emissions.

---

[4]We use projections based on the August 1987 smog episode because this is the only circumstance in which detailed breakdowns of baseline and controlled emissions are presented in the 1997 Plan. August 1987 saw particularly high ozone levels in the South Coast.

Table 2.1

Emission Reduction Targets by Source for NMOG in the South Coast Air Basin[a]

Source	2010 Baseline (tpd) [b]	2010 Controlled (tpd)	Change (tpd)	Percent Change
Stationary Sources	**617.6**	**265.8**	**-351.8**	**-57**
Fuel combustion	10.3	6.0	-4.3	-42
Residential	0.7	0.6	-0.1	-14
Nonresidential	9.6	5.4	-4.2	-44
Solvents use	439.6	157.8	-281.8	-64
Consumer products	91.7	19.9	-71.8	-78
Architectural coatings	97.7	24.4	-73.3	-75
All other	250.2	113.5	-136.7	-55
Petroleum processing, storage, and transfer	45.8	10.3	-35.5	-77
Industrial processes	26.5	13.9	-12.6	-48
Other stationary[c]	95.4	77.8	-17.6	-18
On-Road Mobile Sources	**152.2**	**75.6**	**-76.6**	**-50**
Light-duty vehicles	110.4[d]	55.3[d]	-55.1	-50
Medium-duty vehicles	20.3[d]	10.5[d]	-9.8	-48
Heavy-duty vehicles	21.4[d]	9.8[d]	-11.6	-54
Off-Road Mobile Sources	**133.8**	**59.8**	**-74.0**	**-55**
Off-road vehicles	44.4	21.2	-23.2	-52
Mobile equipment	62.1	19.8	-42.3	-68
Aircraft	18.4	11.9	-6.5	-35
Locomotives	1.3	1.0	-0.3	-23
Commercial boats and ships	3.5	2.7	-0.8	-23
Other	4.1	3.2	-0.9	-22
Total	**903.6**	**401.2**	**-502.4**	**-55**

SOURCE: SCAQMD, 1996, pp. V-C-3 and V-D-3.

[a]Emissions during weather conditions equivalent to the first day of the August 1987 episode.

[b]Based on regulations adopted through September 1996.

[c]Other stationary sources include waste burning, pesticide application, farming operations, and new source review exemptions.

[d]Based on percentage breakdown of on-road mobile source emissions in 1994 SIP (CARB, 1994, p. I-26).

2.3 STRATEGY FOR MEETING EMISSION REDUCTION TARGETS

We now turn to the strategies that are in place to achieve the target emission reductions. We examine the progress that has been made in achieving these reductions in each source category. Tables 2.3 and 2.4 summarize emission reductions for the measures that have been adopted and the gaps that remain. The second column lists the target emission reductions for each source category described in the 1997 Plan (the targets for stationary sources that are not consumer products have been updated using the 1999 Plan). The third column lists the emission

Table 2.2

Emission Reduction Targets by Source for NOx in the South Coast Air Basin[a]

Source	2010 Baseline (tpd)[b]	2010 Controlled (tpd)	Change (tpd)	Percent Change
Stationary Sources	**110.1**	**82.2**	**-27.9**	**-25**
Fuel combustion	81.8	54.3	-27.5	-34
Residential	20.6	13.6	-7.0	-34
Nonresidential	61.2	40.7	-20.5	-34
Solvent use	0.4	0.4	0	0
Consumer products	0	0	0	0
Architectural coatings	0	0	0	0
All other	0.4	0.4	0	0
Petroleum processing, storage, and transfer	2.6	2.6	0	0
Industrial processes	2.7	3.1	0.4	15
Other stationary[c]	22.6	21.8	-0.8	-4
On-Road Mobile Sources	**361.0**	**280.1**	**-80.9**	**-22**
Light-duty vehicles	110.9[d]	102.6[d]	-8.3	-8
Medium-duty vehicles	74.5[d]	55.5[d]	-19.0	-26
Heavy-duty vehicles	175.6[d]	122.0[d]	-53.6	-30
Off-Road Mobile Sources	**267.9**	**162.5**	**-105.4**	**-39**
Off-road vehicles	6.0	5.9	-0.1	-2
Mobile equipment	161.0	94.3	-66.7	-41
Aircraft	24.1	18.3	-5.8	-24
Locomotives	25.7	8.3	-17.4	-68
Commercial boats and ships	39.9	29.0	-10.9	-27
Other	11.2	6.7	-4.5	-40
Total	**739.0**	**524.8**	**-214.2**	**-29**

SOURCE: SCAQMD, 1996, pp. V-C-3 and V-D-3.

[a]Emissions during weather conditions equivalent to the first day of the August 1987 episode.

[b]Based on regulations adopted through September 1996.

[c]Other stationary sources include waste burning, pesticide application, farming operations, and new source review exemptions.

[d]Based on percentage breakdown of on-road mobile source emissions in 2010 baseline and 2010 controlled emissions inventories in 1994 SIP (CARB, 1994, p. I-25).

reductions for the measures adopted to date, and the final column lists the shortfalls from the targets.[5]

---

[5]The combined emission reduction target across all measures (404 tons per day for NMOG) does not equal the target emission reductions in Table 2.1 (502 tpd) for several reasons. First, the numbers in Table 2.3 are for average annual emissions, whereas those in Table 2.1 are for weather conditions corresponding to those of the first day of the August 1987 ozone episode. Second, some measures, such as reductions of NMOG emissions from pesticides, are not included in Table 2.3, because information is not readily available on their emission reductions and, according to CARB, there is no substantial shortfall between the reductions achieved and targeted. Finally, the emission models are complex and fluid, and the data from different sources are not likely to be completely consistent. Table 2.3, though, gives at least a rough feel for the progress that has been made and the gaps that remain. The target NOx reductions in Table 2.2 and 2.4 are much closer.

Table 2.3

Progress Toward Meeting Emission Reduction Targets for NMOG (average annual tons per day)

Measure	Target Reductions by 2010	Projected Reductions by 2010	Projected Relative to Target
Stationary Sources	**261**	**178**	**-83**
Nonconsumer products	184	156	-28
Short- and intermediate-term measures	156	156	0
Long-term measures	28	0	-28
Consumer products	77	16	-61
Consumer products (CP2)	34	16	-18
Long-term measures	43	0	-43
Enhanced vapor recovery	0	6	6
On-Road Mobile Sources	**65**	**24**	**-41**
Light- and medium-duty vehicles (M1, M2, I&M)	19	4	-15
Heavy-duty diesel trucks (M4, M5, M6, M17)	9	6	-3
On-road motorcycles	0	1	1
Heavy-duty gas truck standards adopted in 2000	0	<0.5	<0.5
Urban bus standards	0	0	0
Clean fuels	0	12	12
2007 diesel truck standards	0	1	1
Long-term measures	37	0	-37
Off-Road Mobile Sources	**78**	**65**	**-13**
Off-road diesel equipment (M9, M10)	4	9	5
Off-road gas and LPG equipment (M11, M12)	32	30	-2
Marine vessels (M13)	0	0	0
Aircraft (M15)	3	0	-3
Pleasure craft (M16)	21	25	4
Amendments to small off-road engine standards	0	1	1
Long-term measures	18	0	-18
Total	**404**	**267**	**-137**

SOURCE: Figures for stationary source nonconsumer products taken from SCAQMD, 1999, p. 2-21. All other figures are from CARB, 2001c.
NOTE: "M" denotes mobile source measures identified in CARB's mobile source reduction plan; "I&M" denotes the light-duty vehicle inspection and maintenance (I&M) program.

The situation for NOx looks promising. Adopted measures are expected to achieve more than the required emission reductions for on-road mobile sources, which will offset the shortfall for off-road mobile sources. Emission reductions from the measures currently in place are thereby sufficient to achieve the NOx emission ceiling in the 1997 Plan. Such is not the case for NMOG, however. Even though measures have been adopted that will achieve roughly two-thirds of the emission reductions required, programs that will reduce emissions by another 137 tons per

Table 2.4

Progress Toward Meeting Emission Reduction Targets for NOx (average annual tons per day)

Source	Target Reductions by 2010	Projected Reductions by 2010	Projected Relative to Target
Stationary Sources	**12**	**12**	**0**
Nonconsumer products	12	12	0
Short- and intermediate-term measures	12	12	0
Long-term measures	0	0	0
Consumer products	0	0	0
Consumer products (CP2)	0	0	0
Long-term measures	0	0	0
Enhanced vapor recovery	0	0	0
On-Road Mobile Sources	**85**	**133**	**48**
Light- and medium-duty vehicles (M1, M2, I&M)	17	45	28
Heavy-duty diesel trucks (M4, M5, M6, M17)	62	58	-4
On-road motorcycles	0	<0.5	<0.5
Heavy-duty gas truck standards adopted in 2000	0	1	1
Urban bus standards	0	2	2
Clean fuels	0	12	12
2007 diesel truck standards	0	15	15
Long-term measures	6	0	-6
Off-Road Mobile Sources	**88**	**54**	**-34**
Off-road diesel equipment (M9, M10)	47	42	-5
Off-road gas and LPG equipment (M11, M12)	17	10	-7
Marine vessels (M13)	15	2	-13
Aircraft (M15)	6	0	-6
Pleasure craft (M16)	0	0	0
Amendments to small off-road engine standards	0	<-0.5	<-0.5
Long-term measures	3	0	-3
Total	**185**	**199**	**14**

SOURCE: Figures for stationary source nonconsumer products derived by summing 7.6 tons per day from residential water heaters with 4.2 tpd from small boilers and process heaters (SCAQMD, 1999, pp. 1-5, 2-13). All other figures are from CARB, 2001c.

NOTE: "M" denotes mobile source measures identified in CARB's mobile source reduction plan; "I&M" denotes the light-duty vehicle inspection and maintenance (I&M) program.

day must still be found. This is a very large shortfall. We examine the shortfalls for each of the major source categories in turn.

The largest stationary source NMOG shortfall is in consumer products. The 1997 Plan called for emissions from consumer products to be reduced by 77 tons per day, but measures adopted to date are expected to reduce emissions in 2010 by only 16 tons per day. CARB, which is responsible for regulations on consumer products, does not appear confident that it will be able to adopt regulations to close this gap.[6] A sizable shortfall (28 tons per day) remains for other stationary sources, which are controlled by the SCAQMD. The SCAQMD has identified the sources it plans to target to achieve these reductions but has not begun crafting actual regulations.[7]

There is also a substantial shortfall for on-road mobile sources. CARB had hoped to achieve sizable NMOG reductions (14 tons per day) from a program to buy and scrap older, high-polluting vehicles in the South Coast (measure M1). However, CARB has been unable to obtain the roughly $100 million a year needed to fund the program,[8] and it also has yet to design programs to achieve the 37 tons that were attributed to long-term measures in the 1997 SIP.[9] Indeed, CARB does not seem close to settling on a strategy for finding these additional reductions: There are no publicly available proposals, and even the types of mobile sources that will be targeted have not been identified.

The most progress has been made in achieving the target reductions for off-road mobile sources. Over 80 percent of the target reductions have been achieved; however, 13 tons still remain. In particular, CARB has yet to develop the long-term measures it committed to (in the 1994 SIP) to reduce emissions by 18 tons per day.

The challenge facing California is daunting. Aggressive policies must be adopted to meet the NMOG emission reduction targets that are motivating CARB's vision of a zero emission on-road vehicle fleet. However, even if on-road NMOG emissions were to fall to zero, NMOG emissions from stationary and off-road mobile sources would still exceed the ceiling required to meet ozone standards in the South Coast Air Basin.[10] This dire situation does not necessarily require a zero emission fleet, however. Whether such an approach makes sense depends on whether it is more cost-effective to squeeze the last bit of emission reductions out of the on-road vehicle fleet than it is to reduce emissions from stationary and off-road sources. A fleet with very

---

[6]In a table that summarizes measures to fulfill the ozone SIP commitments, CARB notes that "Development of long-term measures will be coordinated through the Consumer Products Working Group; ARB will reassess feasibility of the commitment as part of Clean Air Plan development in 2001" (CARB, 2001c).

[7]The SCAQMD identified four long-term control measures in its 1999 Plan: solvent cleaning and degreasing operations, 16 tons per day; miscellaneous industrial coatings and solvent operations, 6 tpd; fugitive emissions, 5 tpd; and industrial process operations, 1 tpd (SCAQMD, 1999, pp. 2-14).

[8]See Dixon and Garber, 2001, p. 78, for a discussion of M1 program costs.

[9]These are sometimes called the "black box" reductions.

[10]After the emission reduction target is met, NMOG emissions from on-road mobile sources will total roughly 75 tons per day (see Table 2.1). The shortfall in NMOG emission reductions from stationary and off-road mobile sources is 96 tons per day (see Table 2.3), absent additional emission reduction measures.

low emissions will undoubtedly be required to meet air quality standards in the South Coast Air Basin, but not necessarily a fleet with zero emissions.

2.4 COST-EFFECTIVENESS OF RECENT AND EXPECTED REGULATIONS

Estimates of the cost-effectiveness of alternative methods for reducing emissions are essential to evaluating whether the ZEV program makes sense. This subsection examines the cost-effectiveness of measures recently adopted and currently being considered as ways to reduce NMOG and NOx emissions in the South Coast Air Basin. Tables 2.5 and 2.6 list estimates of the cost-effectiveness of recent or planned stationary and mobile source measures, respectively. The estimates are drawn primarily from analyses by CARB and SCAQMD. Cost-effectiveness is reported in dollars per ton of NMOG plus NOx reduced, and all estimates have been converted to 2001 dollars. To provide a sense of when the costs will be incurred, the tables also show the scheduled implementation dates.

As shown in Table 2.5, the cost-effectiveness of recently adopted CARB regulations ranges from $500 to $6,900 per ton of NMOG reduced (consumer products emit only NMOG, not NOx). The cost of SCAQMD regulations on stationary sources has for the most part been less than $10,000 per ton, but some of the regulations about to be implemented are more expensive, costing up to $25,000 per ton. With only one exception, these regulations have targeted NMOG emissions. Table 2.6 shows that recent regulations on diesel engines have been low cost in terms of dollars per ton of emissions reduced. Regulations on on-road and off-road diesel vehicles have cost less than $800 per ton; recent regulations on gasoline (spark-ignition) vehicles have cost substantially more. The vehicle scrappage (measure M1) and enhanced inspection and maintenance (Smog Check) programs have run to over $30,000 per ton, and regulations on off-road gasoline engines have run to over $20,000 per ton.

Both CARB and the SCAQMD have set guidelines for the cost-effectiveness of their regulations. If the cost exceeds $13,500 per ton of NMOG, the SCAQMD institutes a public review and decision process to seek lower-cost alternatives (SCQAMD, 1999, p. 2-18). Several of the measures the SCAQMD is considering cost more than $13,500 per ton (see lower portion of Table 2.5). It may find less expensive alternatives, but the fact that it is considering these measures suggests it believes it will be unable to achieve its emission reduction targets with measures that cost less than $13,500 per ton. The upper limit set forth in CARB guidance is $22,000 per ton of NMOG plus NOx (CARB, 1998a, p. 60). All the cost estimates developed by CARB (in Tables 2.5 and 2.6) fall below this limit, and the limit is at least contained within the ranges developed by others for the vehicle scrappage and inspection and maintenance programs.

Table 2.5

Cost-Effectiveness of Stationary Source Measures Recently Adopted or Currently Under Consideration (2001 dollars)

Regulation	Pollutant Reduced	Year Implemented	Cost-Effectiveness ($/ton of NMOG+NOx)
Consumer Products[a]			
Aerosol coating products reactivity regulation	NMOG	2000	1,600
Mid-term measures II consumer products regulation	NMOG	2000-2005	900
Mid-term measures consumer products regulation	NMOG	2000-2005	500
Hairspray regulation	NMOG	1997	4,800
Aerosol coating products regulation	NMOG	1995	6,100-6,900
Phase II consumer products regulation	NMOG	1991	Up to 2,400
Antiperspirants and deodorant regulation	NMOG	1989	1,200-2,800
Nonconsumer Products[b]			
Solvent cleaning operations (Rule 1171)	NMOG	1999	1,000
Architectural coatings and cleanup solvents (Rule 1113)	NMOG	2003	25,000
Industrial coatings and solvent operations	NMOG	2002-2003	2,300-8,700
Large solvent and coatings sources	NMOG	2001-2002	6,800-20,400
Large fugitive NMOG sources	NMOG	2001-2003	Up to 20,400
Methanol emissions from hydrogen plant vents (at petroleum refineries)	NMOG	2000	800
Adhesives (Rule 1168)	NMOG	2000	7,800
Solvents usage (Rule 442)	NMOG	2000	2,800
Restaurant operations	NMOG & PM10[c]	2000	10,500
Residential water heaters (Rule 1121)	NOx	1999	12,900

[a]CARB, 2000a.
[b]SCAQMD, 1999.
[c]Particulate matter with a diameter less than 10 microns.

We conclude our discussion of the cost-effectiveness of regulations recently adopted or scheduled to be adopted in the near future with several observations.

First, the variation in cost-effectiveness across the different source categories suggests that it may make sense to pursue reductions in some source categories before pursuing reductions in others. Recent regulations on diesel engines are among the least expensive, and further diesel engine regulation deserves consideration. However, further such regulation may not contribute much to solving the South Coast's ozone problem, because diesel engines emit mainly NOx, and

Table 2.6

Cost-Effectiveness of Mobile Source Measures Recently Adopted or Currently Under Consideration (2001 dollars)

Regulation	Pollutant Reduced	Year Implemented	Cost-Effectiveness ($/ton of NMOG+NOx)
On-Road Mobile Sources			
Scrapping older vehicles (M1) [a]	NMOG & NOx	Insufficient funding	4,000–33,000
Enhanced Smog Check[b]	NMOG & NOx	1996-2002	15,000-31,000
2004 2.5 gm/bhp-hr heavy-duty diesel regulations (M5 and M6)[c]	mainly NOx	2004	400
2007 0.2 gm/bhp-hr heavy-duty diesel regulations[d]	mainly NOx	2007	800
Off-Road Mobile Sources			
Small off-road spark-ignition engine regulations[e]			
0-66 cc handheld engines	NMOG & NOx	2000	600–2,800
>60 cc non-handheld engines	NMOG & NOx	2000, 2004	300–10,000
5-15 hp Tier III relative to Tier II standards	NMOG & NOx	2005	800–20,600
Large off-road spark-ignition engine regulations 3.0 gm/bhp-hr[f]	NMOG & NOx	2001-2004	800–20,600
Off-road diesel regulations (M9 and M10) but 3.0 gm/bhp-hr[g]	mainly NOx	2000-2005	<700

NOTE: gm/bhp-hr = grams per brake horsepower-hour; cc = cubic centimeter; hp = horsepower.
[a]Dixon and Garber, 2001, p.68
[b]Schwartz, 2000, p.16.
[c]CARB, 1998b, p. 64.
[d]CARB, 2001b, p. 49.
[e]CARB, 1998a, p.61.
[f]CARB, 1998c, p.21.
[g]CARB, 1999a, p. 57.

the South Coast is short of NMOG reductions.[11] Regulations on consumer products are also relatively inexpensive. The target emission reductions for consumer products thus appear to make economic sense. Cost per ton will likely rise as the target emission levels are approached, and it may or may not make sense to reduce consumer product emissions beyond the target. We note that few emissions remain from consumer products once the target emission levels have been achieved (20 tons per day, as shown in Table 2.1).

---

[11]The SCAQMD might also consider lowering target NOx emissions and raising target NMOG emissions. As noted above, reductions in NOx emissions can be substituted for reductions in NMOG emissions to some extent.

Regulations on other stationary sources and on gasoline engines are both more expensive. In deciding how to meet air quality standards most cheaply, it should be noted that controls on stationary sources have mainly targeted NMOG, whereas those on gasoline engines have reduced both NMOG and NOx. Thus, the regulations on stationary sources become relatively more attractive when the emission reduction shortfall is for NMOG, not NOx. It should also be noted that NMOG emissions from stationary sources remain large even after the target emission levels have been achieved. For example, the target emission levels for solvents use (outside of consumer products and architectural coatings) is 113.5 tons per day, and the target for other stationary sources is 77.8 tons. These compare with a 75.6-ton target for NMOG emissions from all on-road mobile sources combined (see Table 2.1).

Our second observation about these cost-effectiveness numbers is that it makes no economic sense for the cost-effectiveness guidance to be $13,600 per ton at the SCAQMD and $22,000 per ton at CARB—achieving air quality standards at minimum cost dictates that they be the same. The numbers in Tables 2.5 and 2.6 indicate that the SCAQMD is seriously considering measures with cost-effectiveness near the CARB limit, so the two agencies may be considering measures with similar costs per ton. But even if this is so, the explicit guidance at both agencies should be the same.

Finally, to meet air quality standards in the South Coast Air Basin, it may eventually be necessary to adopt measures that are more expensive than those in Tables 2.5 and 2.6. A complete portfolio of regulations that will achieve air quality standards has not been identified, and as regulations are further tightened, the cost per ton will likely increase. In addition, continued population and economic growth after 2010 will likely entail ever tighter regulations in order to keep overall emissions below a fixed ceiling. Economic efficiency dictates that policymakers be sure that there are no lower-cost alternatives before adopting expensive regulations on a particular source.

3. TECHNOLOGIES FOR MEETING ZEV PROGRAM REQUIREMENTS AND PRODUCTION VOLUME ESTIMATES

This section provides an overview of the vehicle technologies that auto manufacturers may use to meet the ZEV program requirements. (These technologies are described in greater detail in Section 4, where we develop cost estimates.) This section also presents the production volumes we used in our analysis, which are important in determining vehicle unit cost. The production projections are not meant to be precise estimates of the numbers of vehicles that will be produced to meet the program requirements. Rather, they are rough estimates of the number of vehicles that *may* be produced if a particular technology is used in a meaningful way to meet the requirements.

3.1 TECHNOLOGIES FOR MEETING ZEV PROGRAM REQUIREMENTS

As discussed in Subsection 1.1, CARB has created three technology categories within the ZEV program: ZEVs, partial zero emission vehicles (PZEVs), and advanced technology partial zero emission vehicles (ATPZEVs). Large-volume manufacturers must meet a minimum portion of the program requirements with ZEVs and have the option to satisfy other portions with PZEVs and ATPZEVs. The following paragraphs describe the types of vehicles that manufacturers might plausibly produce in each category.

ZEVs

ZEVs can be based on a number of energy storage technologies (e.g., batteries, capacitors, flywheels, fuel cells), but only battery-powered electric vehicles (BPEVs) and fuel-cell vehicles are actively being pursued by the large-volume manufacturers. We first discuss the various types of BPEVs that may be used to meet program requirements, before turning to fuel-cell vehicles.[1]

Battery-Powered Electric Vehicles. Until a few years ago, the only vehicles that manufacturers considered for meeting ZEV program requirements were BPEVs similar in size to many vehicles on the road today and freeway capable. These "full-function EVs" typically have top speeds greater than 65 miles per hour, meet U.S. highway safety standards, include amenities such as air conditioning, and have reasonable acceleration (although often less than that of a comparable internal combustion engine vehicle, or ICEV). The major shortcoming of these vehicles is their range—i.e., the distance they can travel on a single charge. While battery

---

[1]ZEVs may have emissions associated with the production, marketing, and distribution of the fuel they use. For example, power plants generate emissions when producing electricity. These indirect emissions are discussed in Subsection 5.1.

technology has improved during the last decade, barring some further technological breakthrough, it is unlikely that the full-function EVs that manufacturers would produce starting in 2003 would have a real-world range much over 100 miles. The large-volume manufacturers have produced a number of different full-function EVs to date; Appendix A details their characteristics.

Faced with the high production costs of full-function EVs and large projected losses per vehicle, some manufacturers have developed BPEVs that are much smaller than the typical ICEVs they sell in the United States. These "city EVs" have limited top speed, acceleration, and range. They typically seat two passengers and meet highway safety standards but, with a top speed of only 50 to 60 miles per hour, are not designed for regular freeway use. They usually can travel 50 to 60 miles on a single charge. City EVs will soon be available from some automakers (see Appendix A).

"Neighborhood EVs," which are intended for local travel on low-speed-limit routes (especially in and around planned communities), can also be used to satisfy the program requirements. Resembling golf carts (although some have doors), the current models have a range of about 30 miles and are limited by law to a top speed of 25 miles per hour. The National Highway Transportation Safety Administration requires that they have safety features, including lights, mirrors, a windshield, and seat belts, but they are not required to meet the other safety standards of vehicles that can travel at higher speeds. Because CARB believes that neighborhood EVs will not displace many of the miles traveled by vehicles currently on the road, it has severely discounted the number of ZEV credits such vehicles generate starting in model year 2006. We thus think it unlikely that neighborhood EVs will play much of a role in satisfying the ZEV program requirements, except perhaps in the very early years, and we do not consider them further in our analysis.

Direct Hydrogen Fuel-Cell Vehicles. A fuel-cell vehicle is powered by an on-board fuel cell that generates electricity that drives an electric motor. The relatively low-temperature proton exchange membrane (PEM) fuel-cell stack has been singled out by major automakers as the fuel-cell technology choice for autos.[2] PEM fuel-cell vehicles can be fueled with hydrogen, methanol, or even gasoline, but only vehicles fueled with hydrogen qualify as ZEVs. We refer to these as direct hydrogen fuel-cell vehicles (DHFCVs). The other types require a reformer that produces hydrogen from methanol or gasoline in a process that generates carbon monoxide emissions, and there are evaporative emissions associated with gasoline and methanol. It appears that DHFCVs will allow a greater range than BPEVs—the storage capacity of their hydrogen fuel tanks has

_____
[2]PEMs are also referred to as polymer electrolytic membranes or polymer electrolyte membranes, but proton exchange membrane is the most common term.

increased enough that DHFCVs can travel upwards of 180 miles on a single fill-up. (Characteristics of recent prototypes are provided in Subsection 4.1.)

Several of the large-volume manufacturers have developed prototype DHFCVs, but it appears likely that these vehicles will be available only in very small quantities before 2006. Honda recently announced that it will sell DHFCVs to fleets in California in 2003 but expects to make "less than a couple of handfuls" available (Hydrogen & Fuel Cell Letter, 2002). At the Tokyo auto show in October 2001, Toyota announced its intention to start selling a compressed hydrogen fuel-cell vehicle beginning in 2003, with planned sales of 30 to 50 per year (Schreffler, 2001). Ford plans to offer its Focus DHFCV in California starting in 2004. Initial production is likely to be "in the tens" of vehicles, according to Ford (Hydrogen & Fuel Cell Letter, 2002).

The infrastructure to fuel DHFCVs remains a major stumbling block. Various options are being considered, including fueling stations (perhaps part of existing gas stations) with reformers that generate hydrogen from natural gas.

Partial Zero Emission Vehicles

To qualify for PZEV credits, vehicles must meet very stringent emission, durability, and warranty standards. They must be certified to CARB's

- 150,000-mile super ultra low-emission vehicle (SULEV) exhaust emission standard for light-duty vehicles (LDVs);
- zero evaporative emission standard; and
- on-board diagnostics (OBD) system requirements. The OBD system must be able to detect whether the emission standards have been even very slightly exceeded.

In addition, the vehicle warranty on emission control equipment must be extended to 15 years or 150,000 miles, whichever occurs first (CARB, 2002, p. 4).

Current CARB regulations outside the ZEV program require manufacturers to begin producing vehicles that meet SULEV exhaust standards by 2009,[3] so PZEVs do not improve on the tightest tailpipe emission category currently in place. PZEVs must be certified at 150,000 miles, however. Standard SULEVs are certified at 120,000 miles, although their manufacturers can opt to certify them at 150,000 miles in return for greater emission credits.[4] Tighter regulations on evaporative emissions will be phased in between 2003 and 2006, but the zero

[3]The certification standard for ultra low emission vehicles (ULEVs) is 0.04 grams per mile, so once the fleet-average NMOG requirement falls below 0.04, manufacturers must start producing SULEVs. The fleet-average NMOG requirement first falls below 0.04 in 2009 (CARB, 2001g, pp. E-5, E-16).

[4]Manufacturers must demonstrate that their vehicles meet emission requirements before they can be certified for sale in California. Such demonstrations are usually done using protocols that simulate component aging in the lab and extrapolated deterioration rates from road tests on a few demonstration vehicles.

evaporative standard represents a further reduction.[5] The 15-year/150,000-mile warranty extends the comprehensive 3-year/50,000-mile emission warranty currently required in California.[6] Over the years, CARB has created progressively more-stringent vehicle emission control categories. PZEVs are the next step. As we discuss in Section 5, while not zero, NMOG and NOx emissions from PZEVs are exceedingly small.

Advanced Technology Partial Zero Emission Vehicles

An ATPZEV is a PZEV that includes components common to ZEVs—e.g., an advanced battery that is integral to the operation of the power train or an electric power train. The hybrid EVs recently introduced in the United States by Toyota and Honda will qualify as ATPZEVs once they meet the PZEV emission standards.[7] These vehicles run on gasoline and are propelled by a drive train that integrates an electric motor and an internal combustion engine. While other types of ATPZEVs are possible, these gasoline hybrid electric vehicles (GHEVs) appear to be the only ATPZEVs currently being developed by manufacturers.[8] The advantage of a GHEV to the consumer is improved gasoline mileage—GHEV fuel efficiency can be 30 to 50 percent higher than that of a comparably sized standard ICEV.

Because ATPZEVs must meet the same emission standards as PZEVs, they offer no reductions in NMOG or NOx exhaust emissions or evaporative emissions. GHEVs, however, generate less carbon dioxide per mile because of their better fuel efficiency. As we discuss in Section 5, lower fuel use also means lower upstream emissions from the production and distribution of fuel.

3.2 PRODUCTION VOLUMES USED IN COST ANALYSIS

Our ultimate goal in this study was to evaluate the promise of the different technologies for meeting ZEV program requirements. To do this we evaluated the cost-effectiveness of each technology when vehicles are produced in high volume. But we also had to consider the costs incurred when production volumes are lower to evaluate the overall cost-effectiveness of these technologies. We present here the production volumes we used in our analysis. These are not meant to be precise estimates of the numbers of vehicles that will be produced to satisfy

---

[5]The emissions allowed under the zero evaporative standard are very low but not actually zero.

[6]On some emission parts, California requires a 7-year/70,000-mile warranty and the federal government requires an 8-year/80,000-mile warranty, but these warranties are much more limited than California's 3-year/50,000-mile warranty.

[7]Honda's vehicle is the Insight, and Toyota's is the Prius. Ford plans to sell a hybrid version of its Escape sports utility vehicle in 2003.

[8]Grid-connected hybrids can also qualify as ATPZEVs. They can be plugged in and charged; they run on batteries until the batteries run down, at which point the internal combustion engine comes on. No large-volume manufacturers are developing grid-connected hybrids, however.

requirements in California and possibly in the four northeastern states that may also adopt the ZEV program. Rather, they provide rough estimates of the numbers of vehicles that may be produced if a particular technology is used in a meaningful way to meet program requirements.

We developed six different scenarios to capture the range of volumes at which large-volume manufacturers might plausibly produce vehicles to satisfy program requirements. In all scenarios, we assume that manufacturers produce the maximum number of PZEVs allowable. We think this likely because, as discussed in Section 4, the costs of meeting the PZEV standards appear to be modest. Table 3.1 describes the scenarios. In the first two scenarios, manufacturers produce FFEVs to meet their ZEV requirements, but they produce different numbers of ATPZEVs: In scenario 1, they make use of ATPZEVs to the maximum extent possible; in scenario 2, they satisfy only one-half of what they are allowed to meet with ATPZEVs. Toyota and Honda are already selling GHEVs in the United States (although they do not yet meet the PZEV emission and warranty requirements), and Ford has plans to introduce the hybrid version of its Escape sports utility vehicle in 2003. It thus seems almost certain that the large-volume manufacturers will satisfy an important part of the program with GHEVs starting in 2003. Not all manufacturers may choose to produce GHEVs, however, and their popularity with consumers remains uncertain. We thus vary the number of ATPZEVs between the maximum allowed and one-half the maximum.

In scenarios 3 and 4, manufacturers produce city EVs to meet one-half of the ZEV portion of the program. Because city EVs have limited capabilities, we think it unlikely that manufacturers can satisfy more than one-half of the ZEV portion of the program with them. The remaining half is met with full-function EVs.

Manufacturers satisfy the ZEV portion of the program with DHFCVs in scenarios 5 and 6. We concluded above that it is unlikely that DHFCVs will be produced in appreciable volumes before 2006. Thus, in our analysis of DHFCVs, we consider their cost-effectiveness from 2006 on.

For each of these scenarios, we combine projections of the number of vehicles sold per year and the number of ZEV credits generated per vehicle to project the number of vehicles produced each year between 2003 and 2030 (between 2006 and 2030 for DHFCVs). The calculations are quite involved, because the program regulations are complicated (see details in Appendix B). Here we present the resulting range of production volumes for each vehicle type. The lower end of the range for full-function EVs is based on the number in scenario 1 when the program applies only in California; the upper end is based on the number in scenario 2 when the program applies in California and the four northeastern states that may adopt the program.

Table 3.1

Assumptions Used to Develop Production Volume Scenarios

Scenario	Full-Function EV	City EV	ATPZEV	DHFCV
Full-Function EV scenarios				
Scenario 1	Meet full ZEV requirement	0	Maximum possible	0
Scenario 2	Meet full ZEV requirement	0	1/2 maximum possible	0
City EV scenarios				
Scenario 3	Meet 1/2 ZEV requirement	Meet 1/2 ZEV requirement	Maximum possible	0
Scenario 4	Meet 1/2 ZEV requirement	Meet 1/2 ZEV requirement	1/2 maximum possible	0
DHFCV scenarios				
Scenario 5	0	0	Maximum possible	Meet full ZEV requirement
Scenario 6	0	0	1/2 maximum possible	Meet full ZEV requirement

Overall sales volumes, and thus the number of vehicles required in the different scenarios, are 75 percent greater when the program applies in California plus the four northeastern states rather than in California alone. A similar approach is used to determine the number of city EVs, ATPZEVs, and DHFCVs.

The results are presented in Figures 3.1 through 3.5. Production of full-function EVs ranges from 4,000 to 10,000 units in 2003 (see Figure 3.1). In scenario 1, it rises to roughly 90,000 vehicles per year in 2018; the gradual rise thereafter tracks the gradual increase in total vehicle sales over time. Production volume reaches just under 40,000 units in 2018 in scenario 2. To put these numbers in perspective, California sales of vehicles subject to the pure ZEV portion of the program are expected to be roughly 1.0 million in 2003 and 1.7 million in 2012 after heavier light-duty trucks (i.e., LDT2s, which are LDTs with a loaded vehicle weight of >3,750 pounds and gross vehicle weight rating of ≤8,500 pounds) are included in the base to calculate program requirements.[9,10] The figures are roughly 1.75 million and 3.1 million, respectively, if

[9]This is only for sales by the six large-volume automakers. They account for roughly 85 percent of LDV sales in California (Ward's Communications, 1999, pp. 31-36).

[10]CARB's EMFAC2000 emission model (CARB, 1999b) estimates that slightly over 21 million LDVs are on the road in California in 2002.

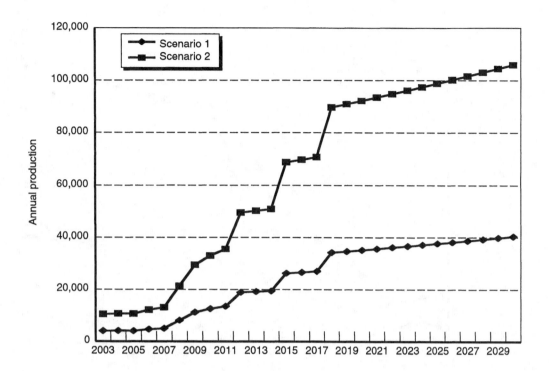

**Figure 3.1—U.S. Production Volume Range for Full-Function EVs
(Based on Scenarios 1 and 2)**

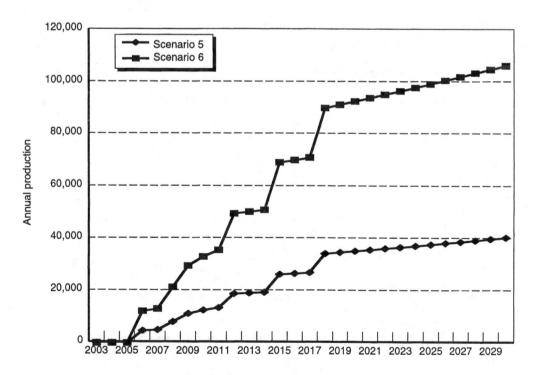

**Figure 3.2—U.S. Production Volume Range for DHFCVs
(Based on Scenarios 5 and 6)**

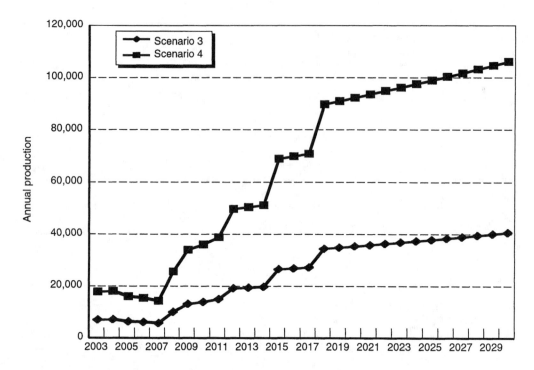

**Figure 3.3—U.S. Production Volume Range for City EVs
(Based on Scenarios 3 and 4)**

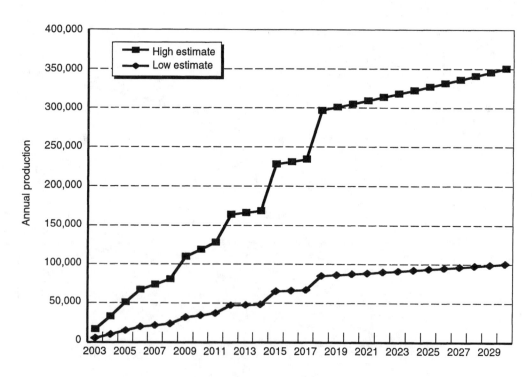

**Figure 3.4—U.S. Production Volume Range for GHEVs
(Based on All Scenarios)**

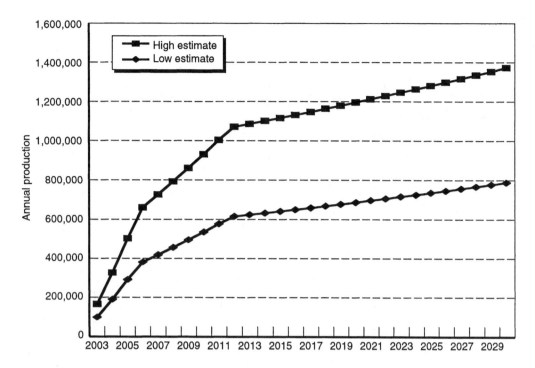

**Figure 3.5—U.S. Production Volume Range for PZEVs
(Based on All Scenarios)**

the four northeastern states are included.[11] The ranges for DHFCVs are the same as those for full-function EVs, except for 2003 to 2006, when production is assumed to be zero (see Figure 3.2). This is because DHFCVs are assumed to generate the same number of ZEV credits as full-function EVs do.

City EV volumes range from 5,000 to 12,000 in 2003 and from roughly 25,000 to 70,000 in 2030 (see Figure 3.3).[12] Note that full-function EVs are also produced in the city EV scenario. In our city EV cost analysis, we take into account the combined volumes of batteries and electric drive components needed to produce these two types of EVs.

The number of ATPZEVs (which we assume are GHEVs) starts low because of phase-in multipliers but then rises quickly (see Figure 3.4). The volume paths for PZEVs are shown in Figure 3.5.[13] The difference between the low- and high-volume scenarios for PZEVs is due

[11]The breakdown of total 1998 sales of passenger cars in the five states that have adopted the program is: California, 57 percent; New York, 29 percent; Massachusetts, 12 percent; Maine, 1 percent; Vermont, 1 percent. Passenger cars account for about 90 percent of total sales of passenger cars plus LDT1s (i.e., LDTs with a loaded vehicle weight of ≤3,750 pounds) (Ward's Communications, 1999, pp. 31-33).

[12]The number of city EVs drops slightly between 2004 and 2007 because of the details of the phase-out of the extended range multiplier and the phase-in of the high-efficiency multiplier.

[13]The PZEV scenarios include PZEV production by intermediate-volume manufacturers, which we assume satisfy their entire program requirements with PZEVs. Based on the number of small-volume

solely to our assuming that the program applies in California only or in California plus the four northeastern states.

---

manufacturers reported by CARB (2000b, p. 8), intermediate- and large-volume manufacturers account for roughly 95 percent of vehicle sales in California. (As noted above, large-volume manufacturers alone account for roughly 85 percent of vehicle sales.)

4. VEHICLE PRODUCTION AND LIFECYCLE COST

This section examines the costs of producing and operating the advanced technology vehicles that manufacturers may build to meet ZEV program requirements. First, we describe the components of battery-powered electric vehicles (BPEVs), direct hydrogen fuel-cell vehicles (DHFCVs), gasoline hybrid electric vehicles (GHEVs), and gasoline vehicles that meet the partial zero emission vehicle (PZEV) emission requirements. Second, we provide an overview of the factors that can cause component costs to fall over time. Third, we discuss the costs of the various components. We estimate costs at the production volumes likely between 2003 and 2007 and at high production volumes. The estimates at high volume are based on well-grounded projections of technological improvements that are plausible given what is known about the technology and about the manufacturing processes possible at high volumes. They are thus the costs that can currently be anticipated after an introductory period during which costs will be higher.

Fourth, we combine the component costs to estimate the incremental cost of each type of advanced technology vehicle over the cost of a gasoline internal combustion engine vehicle (ICEV). The overall social cost of an advanced technology vehicle depends not only on the initial vehicle cost, but also on the cost of operating and maintaining the vehicle over its lifetime. We thus also estimate the incremental lifetime operating costs of each vehicle relative to an ICEV and conclude the section by combining incremental operating cost with incremental vehicle cost to calculate the incremental lifecycle vehicle cost.

4.1 KEY VEHICLE COMPONENTS

We begin by describing the primary components of BPEVs and then turn to the primary components of DHFCVs, GHEVs, and PZEVs.

Battery-Powered Electric Vehicles

The cost increment of a BPEV over a gasoline ICEV comes from the addition of the following basic components:

1. battery modules
2. battery auxiliaries such as tray and straps, harness, and cooling system (modules + auxiliaries = battery pack)
3. electric motor
4. power electronics to control the electric motor
5. electric vehicle transmission

6. on-board (integrated) charger[1]
7. other auxiliaries (e.g., electric power steering, electric brakes).

Cost savings accrue from the exclusion of some components common to conventional vehicles: internal combustion engine and emission control system, fuel tank, and conventional transmission.

Table 4.1 reviews the technical specifications of the BPEVs developed by major manufacturers. Based on this review, we cost out a full-function EV with 30 kWh of battery storage and a 50-kW electric motor between 2003 and 2007. We cost out a city EV with 10 kWh of battery storage and a 25-kW electric motor during the same period. For our high-volume cost estimates, we allow the full-function EV battery pack to vary between 24 and 30 kWh and the city EV battery pack to vary between 8 and 10 kWh to account for plausible improvements in vehicle efficiency. Manufacturers are currently using advanced batteries—nickel metal hydride (NiMH), nickel cadmium (NiCd), and lithium ion (LiIon)—in their full-function EVs, rather than the more conventional lead acid (PbA) batteries. Because NiMH batteries currently appear to be the best suited for automotive use (see discussion in Anderman, Kalhammer, and MacArthur, 2000), we assume they are the advanced battery that auto manufacturers choose for their vehicles.[2] We also run scenarios with PbA batteries because of the possibility that they will be used in some types of vehicles. Due to their low energy density, however, PbA batteries may be difficult to use in vehicles with the ranges we assume here. Their size and weight also create challenges in designing EVs with interior space and handling characteristics comparable to vehicles currently on the road.

Direct Hydrogen Fuel-Cell Vehicles

The cost increment for DHFCVs is due to the following:

1. proton exchange membrane (PEM) fuel-cell system
2. compressed hydrogen tank

---

[1]According to the Pacific Gas and Electric EV glossary, an off-board charger is "a charger with the intelligence and control in the charger stand, not on the vehicle. [An] on-board charger is a charger with the intelligence and control on the vehicle, not in the charger stand." On-board chargers appear to be superior to off-board chargers. According to Korthof (2000), "Integrated charging systems have been developed for electric vehicles by Ford, GM, Renault, Toyota, and Volkswagen. AC propulsion drive systems include an integrated charger rated at 20 kW. This system compares favorably in efficiency, weight, cost, and power quality, and also supports several times more power than comparable isolated conductive chargers or inductive charging systems." An onboard charger also requires some off-board equipment (described in Subsection 4.2).

[2]LiIon batteries have the energy densities needed for vehicles with a 100-mile range, but this technology has shortcomings (see Anderman, Kalhammer, and MacArthur, 2000).

Table 4.1

Specifications of BPEVs

Vehicle	Peak Power (kW)	Energy Storage (kWh)	Battery Type	Range[a] (miles)	Electric Motor Type
Full-Function EVs					
Toyota RAV4-EV	50	27	NiMH	84/78	DC permanent magnet
Nissan Altra EV	62	32	LiIon	95/82	AC
Vehicle modeled in this study	50	24-30	NiMH & PbA	90-110	AC
City EVs					
Toyota e_com	19	~8	NiMH	60	DC permanent magnet
Ford Th!nk City	27	12	NiCd	60	AC
Vehicle modeled in this study	25	8-10	NiMH & PbA	50-60	AC

[a]Numbers separated by a slash show city range with accessories followed by highway range with accessories (see Table A.1 in Appendix A).

3. small (GHEV-sized) battery pack (or other energy storage device)
4. electric motor and transmission
5. power electronics.

Cost savings result from the exclusion of the engine, emissions, control system, transmission, and fuel tank found in an ICEV.

The PEM system is in turn made up of the fuel-cell stack and auxiliary equipment associated with the stack. The auxiliary equipment manages the air, water, and thermal issues related to the system and includes compressors, heat exchangers, a humidification system, safety devices, and a control system (Lipman, 1999b, p. 12). The fuel-cell stack is made up of the following components:

1. membrane electrode assembly (MEA)
   a. cathode
   b. anode
   c. electrolyte/membrane
   d. cathode and anode catalyst (platinum); and
2. bipolar (separator) plates.

A fuel-cell stack consists of many fuel cells. The structure of a single fuel cell is illustrated in Figure 4.1.

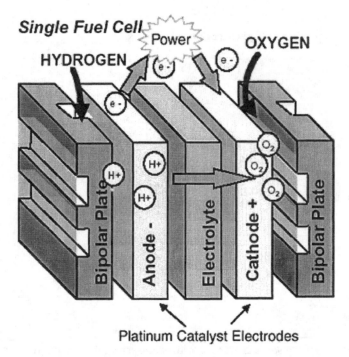

SOURCE: Chalk, 2002.

Figure 4.1—Single PEM Fuel Cell

Ballard Power Systems, Incorporated, a leader in the development of PEM fuel cells, describes the key components and principles of operation of a fuel cell as follows:

> A single fuel cell consists of a membrane electrode assembly and two flow field [bipolar] plates. Single cells are combined into a fuel cell stack to produce the desired level of electrical power. Each membrane electrode assembly consists of two electrodes (anode and cathode) with a thin layer of catalyst, bonded to either side of a proton exchange membrane (PEM). Gases (hydrogen and air) are supplied to the electrodes on either side of the PEM through channels formed in the flow field [bipolar] plates. Hydrogen flows through the channels to the anode where the platinum catalyst promotes its separation into protons and electrons. On the opposite side of the PEM, air flows through the channels to the cathode where oxygen in the air attracts the hydrogen protons through the PEM. The electrons are captured as useful electricity through an external circuit and combine with the protons and oxygen to produce water vapor on the cathode side (Ballard, 2002).

The feasibility of compressed hydrogen fuel tanks has been shown in a number of prototype vehicles. Table 4.2 lists the specifications of some of the prototype DHFCVs that have been produced by various automakers. Manufacturers are experimenting with different energy storage devices and different balances between battery and fuel-cell size. Some are using NiMH batteries to store energy and provide power; others are using supercapacitors. Some designs have a large fuel cell (over 60 kW) and a modest battery (25 kW); others have the reverse. For our

Table 4.2

Specifications of Prototype DHFCVs

Make/Model	Capacity and Pressure of Hydrogen Tank	Top Speed (mph)	Range (miles)	Power Plant (kW)	Battery Power (kW)	Electric Motor Type and Power (kW)
Honda	5,000 psi, 130-liter tank	87	185+	78	n.s	60 kW AC
Toyota Highlander FCHV-4	3,700 psi	95	155	90	n.s.	80 kW DC
Ford Focus FCV	5,000 psi, two 41-liter tanks	80	155	80	n.s.	67 kW AC
Hyundai Santa Fe FCEV	5,000 psi	77	100+	75	n.s.	65 kW
Vehicle modeled in this study						
Option 1	5,000 psi	n.s.	150+	25	60	75 kW
Option 2	5,000 psi	n.s.	150+	60	25	75 kW

NOTE: n.s. = not specified.

near-term projections, we assume a 25-kW fuel cell because high fuel-cell costs in the near term suggest that it will be cheaper to keep the fuel cell small. For volume production, we examine the costs of both vehicles with relatively large and relatively small fuel cells (see options 1 and 2 in Table 4.2). Range should be over 150 miles with the fuel tank we specify in our analysis.

Gasoline Hybrid Electric Vehicles

The cost increment associated with a GHEV comes from the addition of the following components:

1. small battery pack
2. electric motor
3. power electronics (e.g., motor controller).

Cost savings over a conventional ICEV result from the smaller internal combustion engine needed to power a GHEV.

Table 4.3 lists the specifications of GHEVs that have been introduced in the United States or are scheduled to be introduced in the near future. Based on these figures, we cost out a GHEV with an NiMH battery producing 25 kW of power, an electric motor of 33 kW, and an internal combustion engine that is 33 kW smaller than it would need to be in a conventional ICEV.

Gasoline Partial Zero Emission Vehicles

A gasoline PZEV must meet very stringent emission, durability, and warranty standards. The vehicles must be certified to CARB's

- 150,000-mile super ultra low emission vehicle (SULEV) exhaust emission standard for light-duty vehicles (LDVs),

Table 4.3

Specifications of Typical GHEVs

Model	ICE Peak Motor Power (kW)	Battery Type	Battery Power (kW)	Energy Storage (kWh)	Electric Motor
Honda Insight	50	NiMH	~20	0.94	10 kW DC permanent magnet
Honda Civic GHEV	63	NiMH	9.7	0.86	10 kW DC permanent magnet
Toyota Prius	52	NiMH	25	1.4	33 kW DC permanent magnet
Vehicle modeled in this study	52	NiMH	25	not specified	33 kW DC permanent magnet

NOTE: ICE = internal combustion engine.

- zero evaporative emission standard, and
- on-board diagnostics (OBD) system requirements. The OBD system must be able to detect emissions that exceed the standards even very slightly.

In addition, the vehicle warranty on emission control equipment must be extended to 15 years or 150,000 miles, whichever occurs first.

As is discussed in Subsection 4.4, major progress has been made in designing vehicles that meet PZEV standards. The required emission control system is similar to that in vehicles on the road today.

4.2 OVERVIEW OF FACTORS THAT CAUSE COSTS TO DECLINE

Before examining the costs of vehicle components, it is useful to review the various factors that cause costs to fall. Economists and engineers generally agree that three factors, or processes, can lead to reductions in unit component cost: production scale economies, learning, and technological advances.

Production Scale Economies

Higher rates of production per period usually cause the unit cost of production to drop. They can mean discounts on bulk purchases of input materials and components and more efficient usage of machinery and labor. They can also justify the increased use of automated manufacturing equipment, which can further reduce unit costs. In addition, higher production volumes can justify changes in product designs that reduce production costs. For example, it may not be cost-effective to design application-specific integrated circuits until production volumes are high.

Predictions of the cost savings due to scale economies are based on what is known about the product design and technological improvements that can be incorporated at high production volumes. They are also based on what is known about the manufacturing processes possible at high volumes.

Learning

Costs may drop over time as learning-by-doing improves the efficiency of the manufacturing process. These reductions are in addition to those solely caused by an increase in production volume. Refinements or wholesale changes in the production process may be discovered based on accumulated production experience. For example, production experience may lead to a better way to combine the various steps of the production process or to a more efficient mix of capital and labor.

Technological Advances

Improvements in the technical design of a component can reduce its unit cost. For example, advances that allow reductions in the amount of nickel per kilowatt-hour storage capacity will reduce NiMH battery costs.

In our analysis of costs, we restrict our attention to component designs that appear feasible given what is known about the technology. While a specific design does not need to currently exist, there should be a clear path to it that appears feasible. Technology breakthroughs that cannot currently be foreseen may indeed happen in the future, but it does not make sense to base policy decisions on the presumption that they will.

How Cost Reduction Factors Enter Our Analysis

We carefully consider the effects of production scale on our estimates, and the high-volume estimates are based on knowledge about the types of production processes that are feasible at high volumes for the products considered. Manufacturing experience may result in additional cost reductions, and experience in many settings suggests that costs do fall substantially with manufacturing experience. But this is not guaranteed. It could be that the manufacturing process envisioned when developing the high-volume estimate does not work out as well as conceived. Component costs thus may indeed fall below those currently predicted at high production volume, but it is also possible that even with additional manufacturing experience, unanticipated problems may mean that the high-volume predictions are not realized.

In the analysis below, we provide some evaluation of the opportunities for additional cost reductions through learning. Generally, we believe that the high-volume cost estimates in the studies we reviewed have squeezed costs down about as far as possible given the material requirements of the product. Thus, we think it unlikely that costs will fall below the high-volume cost estimates here absent technological advances that currently cannot be foreseen.

Our analysis combines estimates of the relation between production volume and cost and expected production volume between 2003 and 2007 to predict component costs during the first five years of the program. We think it likely that these cost projections are reasonably realistic for what will be observed in the near term. We then predict costs at high production volumes. Forecasting technological advances is very difficult beyond 10 years or so. We thus think it appropriate to interpret our high-volume cost estimates as the lowest costs that can be expected over the next 10 years or so given what is currently known about advanced vehicle technologies and manufacturing processes.

4.3 COMPONENT COSTS OF BATTERY-POWERED ELECTRIC VEHICLES AND GASOLINE HYBRID ELECTRIC VEHICLES

This subsection synthesizes the results of existing studies to estimate the costs of the components in BPEVs and GHEVs. Subsequent subsections address the costs of DHFCV and PZEV components.

Battery Modules and Auxiliaries

NiMH Battery Modules. CARB's Battery Technology Advisory Panel (BTAP) assessed the cost, performance, and availability of batteries for EVs and issued a report in 2000 (Anderman, Kalhammer, and MacArthur, 2000). It based its analysis on data from battery manufacturers, automakers, existing research reports, and its own analysis. It concluded:

> From the cost projections of manufacturers and some carmakers, battery module specific costs of at least \$350/kWh, \$300/kWh and \$225-250/kWh can be estimated for production volumes of about 10k, 20k and 100k battery packs per year, respectively. To the module costs, at least \$1,200 per battery pack (perhaps half of that sum in true mass production) has to be added for the other major components of a complete EV-battery, which include the required electrical and thermal management systems (Anderman, Kalhammer, and MacArthur, 2000, p. v).

The Panel assumed only incremental technological improvements (p. vi), so it based its projections on what is known about the technology today. The declines in unit cost with production volume are due to "economies of scale that result from discounts on bulk purchases of materials and components, higher efficiencies in the use of labor and equipment, and especially, use of custom-designed automated manufacturing equipment with high production rates and product yields" (p. 23). The Panel believes that further reductions in module costs below $225 to $250 per kilowatt-hour are possible with sustained production on a true mass production scale (100,000 or more battery packs per year).[3] These reductions would be due to additional automation, incremental improvements in battery design, and process technology refinement based on accumulated production experience. The Panel did not predict how low costs could go, but it concluded that reductions in costs below $225 to $250 per kilowatt-hour would be possible only if material costs declined significantly (p. 24).[4] We thus conclude that additional learning by experience will produce only limited cost savings given what is currently known about the physics of battery development.

Figure 4.2 displays a curve fit to approximate the BTAP estimates. Data presented by Panasonic EV at the 1999 UC Davis conference are also included in the figure. In our analysis of EV costs at high production volumes, we allow NiMH module costs to vary between $225 and $250 per kilowatt-hour.

Lead Acid Battery Modules. The BTAP cites a range of $100 to $150 per kilowatt-hour for the production costs of lead acid (PbA) batteries at volumes of 10,000 to 25,000 packs per year (Anderman, Kalhammer, and MacArthur, 2000, p. iv). Lipman (1999b) estimates that PbA battery costs are between $107 and $113 per kilowatt-hour for high volumes (>100,000 full-function battery packs per year). Data provided to AC Propulsion and CARB by a major EV battery manufacturer (the name is not publicly available) also provide a half-dozen data points. We used these data to fit a relationship between cost and production volume (see Figure 4.3, where the data represent the cost of a 60-Ah battery module).

In the early years of the ZEV program, manufacturers may produce 5,000 or fewer full-function EVs. If the entire 5,000 full-function EVs were produced with PbA batteries, annual battery production would be 150,000 kWh. The resulting battery cost from Figure 4.3 would be roughly $150 per kilowatt-hour. Figure 4.3 suggests that $100 per kilowatt-hour is a plausible high-volume cost for PbA battery modules.

Table 4.4 gives battery costs for two production scenarios between 2003 and 2007 and for high-volume production. The high full-function EV scenario (2003-2007) accounts for an average annual production of 16,600 vehicles between 2003 and 2007; the low full-function EV

---

[3]Note that there are multiple battery modules in a battery pack.

[4]The Panel concluded that major advances or breakthroughs that could reduce battery costs were unlikely for the next six to eight years—that is, through 2006 or 2008 (Anderman, Kalhammer, and MacArthur, 2000, p. vi).

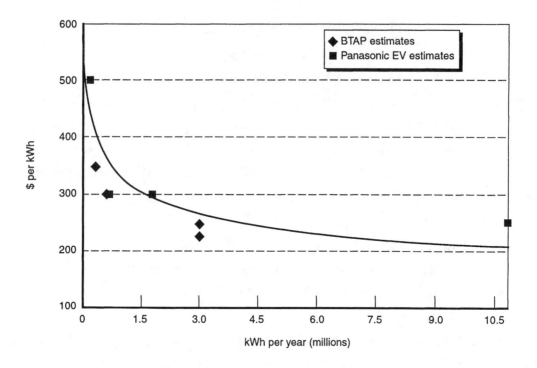

Figure 4.2—NiMH Module Cost as a Function of Production Quantity

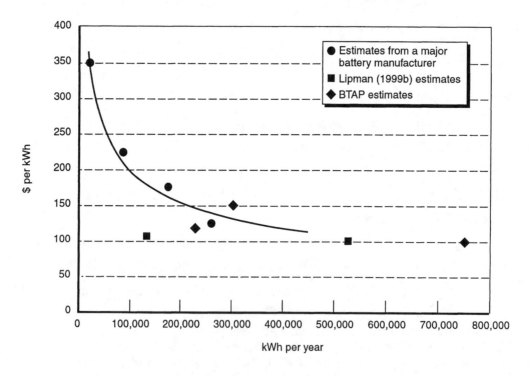

Figure 4.3—PbA Module Cost as a Function of Production Quantity

Table 4.4

EV Battery Module Costs
($ per kWh)

Battery Type	High Full-Function EV Scenario 2003-2007	Low Full-Function EV Scenario 2003-2007	Cost in High-Volume Production
PbA	103	174	100
NiMH	356	465	225-250

scenario accounts for an average annual production of 4,700 vehicles (see scenarios 1 and 2 in Section 3).

NiMH Batteries for GHEVs. Batteries for GHEVs require a higher power density than those for full-function EVs and city EVs. Consequently, these batteries are costed as a function of power. NiMH GHEV batteries provide approximately four times the power density of NiMH EV batteries. (Current designs from Sanyo and Ovonic claim to achieve a power density of 1000 watts per kilogram of battery weight.) According to a recent report, the cost to auto manufacturers is $500 to $2,000 per GHEV battery pack, with a life expectancy of five to eight years (PRNewswire, 2001). Given the power ratings on the current fleet of GHEVs, that equates to roughly a range of $20 to $80 per kilowatt. Lipman, Delucchi, and Friedman (2000) develop a range of $22 to $31 per kilowatt with future manufacturing experience taken into account. We use $20 per kilowatt as a high-volume cost for a NiMH GHEV battery.

Battery Auxiliaries. Battery auxiliaries must be added to the battery modules to create a complete pack. They include (1) a battery tray and straps, (2) an electrical wire harness, (3) bus bars and terminals, and, perhaps, (4) a cooling/ventilation system. Delucchi et al. (2000) provide information on the cost of battery auxiliaries. Based on this study, we put the costs of auxiliaries at $14 to $19 per kilowatt-hour for a PbA battery pack and $8 to $11 per kilowatt-hour for a NiMH pack between 2003 and 2007. As quoted above, the BTAP puts the cost for a 30-kWh full-function EV pack at $1,200 but believes that this figure could be halved in high-volume production. We therefore interpret the Panel estimates to be between $600 and $1,200 depending on production volume, equivalent to $20 to $40 per kilowatt-hour of pack storage capacity. We combine these findings to estimate the costs of battery auxiliaries as $8 to $40 per kilowatt-hour for NiMH packs and $14 to $19 for PbA packs; these estimates hold for the different types of EVs that may be used to satisfy ZEV program requirements. The lower ends of the ranges are used to develop high-volume cost increments for the entire EV (see Table 4.5). While not insignificant, the costs of battery auxiliaries are much less than the costs of battery modules.

Table 4.5

EV Battery Auxiliary Costs
($ per kWh)

Type	2003-2007	Volume Production
PbA auxiliaries	14-19	14
NiMH auxiliaries	8-40	8

Motors and Controllers

Two types of motors and motor controllers are used by major original equipment manufacturers (OEMs) for BPEVs: AC and DC motor/motor controllers. It is accepted in the literature that motor costs are a function of peak power requirements but that motor controller costs are not as closely tied to the motor's peak power. Generally, AC motors are less expensive than DC motors, but the opposite holds for motor controllers (since DC controllers require fewer power switching modules).

Our estimates of motor and motor controller costs, as well as the estimates discussed below for EV transmission costs, are largely based on the work of Lipman (1999a) and Lipman, Delucchi, and Friedman (2000). A brief description of the methods used in these analyses is warranted.

Lipman estimates component costs for different production volumes by estimating parameters for material costs, costs of adding values to materials (e.g., labor and equipment costs), and manufacturer profit (Lipman, 1999a, p. 44). The parameters are estimated in part using data provided by component manufacturers. His estimates are also based on work done at Argonnne National Laboratory (ANL). ANL has extensively analyzed the relationships among material, labor costs, production equipment, and other costs in conventional vehicles produced at high volumes, and Lipman uses these multipliers to help estimate EV component costs at high volumes. He develops high, central, and low estimates of cost at each production volume, which in the case of motor controllers reflect "different assumptions about the degree to which costs of key motor controller components will be reduced by production volume" (Lipman, 1999a, p. 44). We use Lipman's central cost estimates. He estimates production costs for 2,000, 20,000, and 200,000 units per year. We use curves fit to these data to predict costs between 2003 and 2007. For high-volume production, we use the equations to predict costs when annual production volume is 500,000 units.

Lipman's estimates clearly capture technological scale effects, and his projections start with detailed descriptions of the parts and costs of current technologies. Thus, his cost estimates appear to be based on technological improvements that are reasonably likely. His high-volume cost projections consider ratios between materials cost, overhead, and other costs that are based on volume production of conventional vehicles. Thus, he is incorporating reductions in manufacturing costs that can be expected with increased volume based on past experience. Additional cost reductions may be possible as manufacturers gain experience with the production process. On the whole, however, Lipman's description of his approach suggests to us that his analysis contains enough optimistic assumptions that make it unlikely that learning will cause costs to fall beyond our high-volume cost estimates for the next 10 years or so absent unanticipated technological improvements.[5]

AC Induction Motor and Controller. Following Lipman, Delucchi, and Friedman (2000), we assume that the cost of AC induction motors is not sensitive to further increases in production volume and is $10.80 per kilowatt-hour of rated peak power from 2003 to 2007. Lipman argues that existing production of AC motors is large enough that cost is steady across production volumes: The tooling for these motors is in place, and they are produced on flexible-flow production lines (Lipman, 1999a, p. 45).

We follow Delucchi et al. (2000), who cite the data from Lipman and SatCon Technology Corporation in developing costs for AC controllers for several production quantities. The cost estimates when production volume is 200,000 units a year are based on data from SatCon. SatCon has recently received funding from the Department of Energy to develop EV motor and controller components. SatCon seeks to reduce controller manufacturing costs by selecting low-cost material, integrating subsystems to reduce parts counts, and utilizing low-cost production techniques (Lipman, 1999a, p. 43). SatCon's projections at 10,000 to 200,000 units per year represent substantial cost reductions over costs today (from costs over $3,300 today to $500 to $700 for a 50-kW-peak motor). In principle, additional cost reductions are possible with manufacturing experience, but it seems likely that in developing its cost targets, SatCon has squeezed margins above material costs very hard.

Delucchi et al. (2000) include a cost component that is a function of power rating and a component that reflects internal elements of the controller that are not necessarily cost-sensitive to the power rating but are cost-sensitive to production volume increases. Based on these data, we fit the following equation for AC controller costs:

$7.10 to $9.00 per $kW_{motor\text{-}peak}$ + y, where
$y = 149{,}255x^{-0.51}$ and x is the annual unit production volume.

---

[5]For example, his central-cost projections for controllers are based primarily on *target* data for costs from SatCon Technology Corporation (Lipman, 1999a, p. 47).

DC Permanent Magnet Motor and Controller. Lipman, Delucchi, and Friedman (2000) surveyed the literature on DC permanent magnet motors (including Cuenca, Gaines, and Vyas, 1999) and determined costs on the basis of both peak kilowatt rating and production volume. His estimates run from $12.51 to $29.69 per peak kilowatt depending on production volume, not including additional power-invariant costs. Lipman adds $779 for production volumes as low as 2,000 per year and $88 for production volumes around 20,000 per year.[6] His estimates are shown in Figure 4.4 for three different production volumes; we can interpolate for production quantities between these levels.

The material costs used in Lipman's estimates of DC permanent magnet motors are generally based on Cuenca, Gaines, and Vyas's analyses (Lipman, 1999a, p. 45). Cuenca and his colleagues at ANL have developed detailed material costs and high-volume manufacturing costs for permanent magnet motors. At high volume, material costs account for 60 to 70 percent of overall component cost per unit (Lipman, 1999a, p. 75). Thus, absent technological changes in the design, there is not much room for reductions in manufacturing costs through manufacturing experience.

Our estimates for DC permanent magnet controller costs are drawn from the same sources as the AC controller costs discussed above. The resulting equation is

$4.80 to $6.90 per $kW_{motor-peak} + y$

where $y = 162,123x^{-0.53}$, and x is the annual unit production volume.

Motor and Controller Costs Used in Our Analysis. For 2003 to 2007, we combine the above equations with estimates of production volume from Subsection 3.2 to predict motor and controller costs.

Cuenca, Gaines, and Vyas (1999) provide a starting point for estimates of high-volume production costs. As shown in Table 4.6, the materials in their DC system (i.e., motor and controller) cost $1,190 to $1,360, and the materials in the AC system run from $1,316 to $1,516 for the size of the motor indicated. For the 50-kW motor used in our analysis, these estimates translate to between roughly $1,200 and $1,500 for the motor and controller costs combined. Since these estimates were made, the costs and sizes of motor controller parts have declined. GM reported in 1999 that it had cut the cost of its EV1 drivetrain (motor and motor controller) in half, reduced its size by half, and cut the number of parts by one-third as part of its "Generation II" design. At one point, GM's Advanced Technology Vehicle center was working on "Generation

---

[6]These power-invariant costs can be approximated with the function $687,000x^{-0.90}$, where x is the annual production volume.

-45-

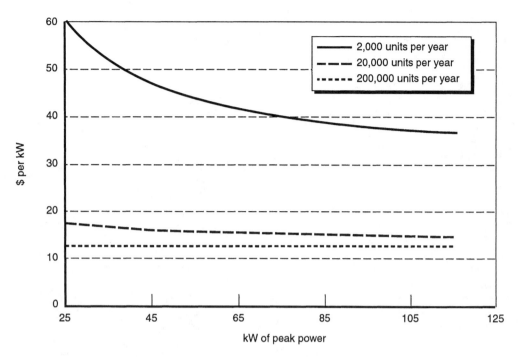

SOURCE: Based on Lipman, Delucchi, and Friedman, 2000.

Figure 4.4—DC Motor Cost as a Function of Production Quantity

III" drive components that were to be half again the cost of the Generation II components (*EV World*, 1999). CARB notes that for the GM Precept, equipped with Generation III drivetrain components, a key component of the power inverter (motor controller) module is one-sixth the size of the one used in the EV1. Honda has published similar statements in regard to electric motors and controllers. Therefore, it seems likely that the material costs in Cuenca, Gaines, and Vyas, 1999, are high given the technological improvements that have already occurred since their study.

We extrapolate from the relationships between production volume and cost above to predict motor and controller costs at high volume. The relationship between combined motor and controller costs and production volume for a 50-kW motor is presented in Figure 4.5. The combined cost at 500,000 units lies between $725 and $1,200 (compared to $1,200 to $1,500 in Cuenca, Gaines, and Vyas, 1999). We use this range in our high-volume production scenarios. As a city EV system is rated at half the power of a full-function EV system, we use a figure that is roughly half these estimates for the city EV motor and controller in high-volume production. The motor in the DHFCV that we cost out is rated at 75 kW. We thus increase the range estimated for a 50-kW system by 50 percent: The resulting range is $1,100 to $1,800.

Table 4.6

Estimate of High-Volume Costs for 53-kW DC Motor, 67-kW AC Motor, and 70-kW DC and AC Motor Controllers

($)

Cost Driver	DC System	AC System
Motor		
Magnets	175	0
Other materials	215	291
Assembly and test	97-167	125-194
Total mfg. cost	487-557	416-485
Gross margin (at 20)	97-111	82-97
Total cost	584-668	498-582
Cost per kW	11-12.6	7.4-8.7
Motor Controller		
Switching modules	200-300	400-600
Other components	500	500
Assembly costs	100	100
Total mfg. cost	800-900	1000-1200
Supplier margin (at 20%)	160-180	200-240
Total OEM cost range	960-1080	1200-1440
Cost per kW	13.7-15.4	17.1-20.6
Total Motor and Controller ($/kW)	24.7-28.0	24.5-29.3

SOURCE: Cuenca, Gaines, and Vyas, 1999.

Summary of Motor and Controller Costs for BPEVs. Table 4.7 summarizes our estimates of the average motor and motor controller costs. The second column predicts costs between 2003 and 2007 for the high-volume scenarios (see Section 3); the third column predicts cost for the same period for the low-volume scenarios. The final column lists high-volume production costs.

Learning through manufacturing experience could conceivably reduce costs below the volume production costs listed here. However, we have been aggressive about continuing to reduce costs as volume increases (production scale economies) and thus think it unlikely that learning will reduce costs further over the next 10 years or so given what is known about the technology and manufacturing processes.

Motor and Controller Costs for GHEVs. Energy and Environmental Analysis estimates the volume production costs for the Toyota Prius (Duleep, 1998). It put the motor controller at $704 and its 33-kW electric motor at $147. Accordingly, we use the sum ($851) as our upper bound for high-volume cost for a GHEV motor and motor controller. GHEV motor and controller costs should not exceed those for a full-function EV in the long run. We thus set the lower bound of the range at $725.

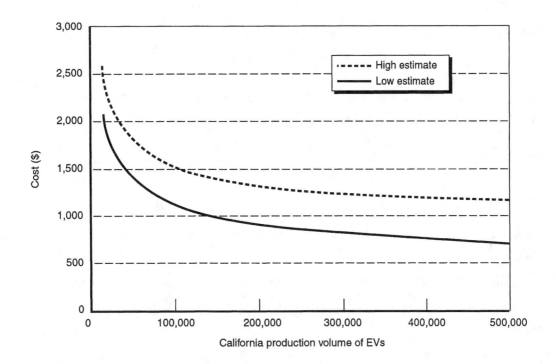

Figure 4.5—Combined Cost of 50-kW Motor and Motor Controller at Different Production Volumes

Table 4.7

**Motor and Motor Controller Costs for Full-Function EVs and City EVs
($ per motor and controller)**

Type	High-Volume Scenarios 2003-2007	Low-Volume Scenarios 2003-2007	High-Volume Production
50-kW Full-Function EV			
AC	2,000-2,100	3,000-3,100	725-1200
DC	2,100-2,500	3,500-3,850	725-1200
25-kW City EV			
AC	1,550-1,600	3,000-3,100	450-550
DC	1,550-1,800	3,500-3,850	450-550

EV Transmission and Other EV Auxiliaries

Lipman (1999b) reports data from Unique Mobility on the cost of a single-speed EV transaxle for volumes of 2,000, 20,000, and 200,000 units per year.[7] Based on those data, we fit the following equation:

---

[7]Unique Mobility considers itself to be a "world recognized technology leader in the development and manufacture of energy efficient, power dense electric motors, generators and electronic inverters aimed at high growth and emerging markets" (http://www.uqm.com).

$$\text{cost per kW}_{\text{motor-peak}} = 315x^{-0.30},$$

where x is the annual unit production volume.

For EV auxiliaries (including motors to drive compressors for steering and braking systems), we use the data in Delucchi et al., 2000 (p. 40), to fit the function

$$\text{cost per kW}_{\text{motor-peak}} = 7.2x^{-0.18} + 0.45x^{-0.20},$$

where x is the annual unit production volume.

The costs of these auxiliaries are small, amounting to roughly \$2 per kilowatt of peak motor power when annual production is 2,000 and \$0.85 per kilowatt when annual production is 200,000 units.

The high-volume estimate for the EV transmission and auxiliaries is based on predictions of these equations when volume is 500,000 units per year. It comes to \$7 per kilowatt of peak motor power.

Integrated Charger

For integrated chargers, we use estimates obtained from AC Propulsion (Brooks and Gage, 2001): \$1,338, \$474, and under \$300 for vehicle volumes of 100, 1,000, and 10,000, respectively, for a conductive system.[8] These cost figures fit the following curve:

$$\text{cost} = 5417x^{-0.32},$$

where x is the annual unit production volume.

This curve predicts a cost of \$284 at a volume of 10,000, and we used it to project integrated charger costs between 2003 and 2007.

AC Propulsion had earlier (in 2000) provided Delucchi et al. with integrated charger cost estimates (Delucchi et al., 2000, p. 39). The costs were \$800 per vehicle when production ranged from 5,000 to 10,000 units per year. The 2001 estimates are substantially below this and may reflect technical improvements or projected improvement in manufacturing processes. We have not been able to investigate the reason for the decrease. Data are not available on high-volume production costs. The substantial decrease in estimated costs between 2000 and 2001 may indicate that costs will continue to decrease in the future. But they may also mean that the 2001 estimates have squeezed costs hard (AC Propulsion is a strong advocate of EVs). We conclude that it is reasonable to use \$250 to \$300 for high-volume production costs.

---

[8]In June 2001, CARB adopted on-board conductive charging as the standardized charging system for EVs in California, with implementation to be in 2006.

Savings From Exclusion of Internal Combustion Engine Vehicle Components

It is important to consider the savings that accrue because several key components of ICEVs are not needed in EVs. In Cuenca, Gaines, and Vyas, 1999, manufacturing costs represent 50 percent of the manufacturer's suggested retail price (MSRP). The engine, transmission, fuel, and exhaust system represent 22 percent of the manufacturing cost (or 11 percent of the MSRP) for a subcompact, and 20 percent (10 percent of MSRP) for a minivan.[9] Based on these percentages and a review of current MSRPs for subcompacts and minivans, we put costs uniquely associated with ICEVs at $16 to $18 per kilowatt of rated engine power when nonmanufacturing costs are excluded and $32 to $36 once nonmanufacturing costs are included.

The cost of ICEV components not required for an EV may well increase. Auxiliary power requirements are expected to rise to the point that existing auxiliary power systems will have to be improved for vehicles with internal combustion engines as the sole power plant. One published estimate puts the cost increment (for more powerful batteries, alternators, and wire harness) at $1,000 (Sullivan, 2001).

Currently, passenger cars operate between 1 and 2 kW of auxiliaries, but this amount is rising as amenities such as cell phones, entertainment systems, navigation systems, electric steering, electric brakes, electric suspension, and electric valves are added. Existing 14-volt generators on ICEVs are limited to generating around 2 kW maximum. All together, nonpropulsion-related power needs could rise to between 5 and 10 kW or more (see Table 4.8), which is comparable to the power needs of a large home.

Cars with 42-volt electrical systems, designed for increased loads, may arrive in the next several years. These systems entail added costs for a larger battery, more robust wiring, and advanced switching components. EVs and hybrid EVs are inherently capable of producing large amounts of electrical power, so it is possible that the cost differential between conventional ICEVs and EVs and hybrid EVs may narrow as ICEV electrical systems are upgraded. We do not make an adjustment for this effect in our analysis. First, we have not been able to examine the basis for the $1,000 estimate above. Second, the extra power requirements will presumably also require more battery storage (and higher costs) to maintain the same vehicle range.

Our discussion so far concerns the costs of ICEV components that are not required in EVs. But our cost estimates of GHEVs must also include the cost savings due to the smaller internal combustion engines that are feasible in hybrids. For example, the Ford Escape GHEV will have a 4-cylinder internal combustion engine that performs like a conventional 6-cylinder power plant. The increment in MSRP from the 4-cylinder to the 6-cylinder is $1,400 according to Ford's price

---

[9]In Cuenca, Gaines, and Vyas, 1999, a subcompact has a 4-cylinder engine and a 5-speed transmission; and a minivan has a 6-cylinder engine and a 4-speed transmission.

Table 4.8

Breakdown of Current and Potential Power Needs of an ICEV

Load	Peak (watts)	Average
Electromechanical valves (6 cyl @ 6,000 rpm)	2,400	800
Water pump	300	300
Engine cooling fan	800	300
Power steering (all elec.)	1,000	100
Heated windshield	2,500	200
Catalytic converter pre-heat	3,000	60
Active suspension	12,000	360
Comm/nav/entertainment	--	100
Total	--	2,220

SOURCE: Kassakian, 2000.

guides. We assume that the MSRP increment is twice the increment of the engine cost, putting the reduction in engine cost at $700.

4.4 COMPONENT COSTS OF DIRECT HYDROGEN FUEL-CELL VEHICLES

In this subsection, we first provide an overview of studies on projected proton exchange membrane (PEM) fuel-cell system costs. We then examine the costs of key fuel-cell components and their cost drivers.

Overview of Recent Studies on PEM Fuel-Cell Systems

The cost today of fuel cells is very high. Current estimates for a hydrogen/air fuel-cell system run from $500 (Arthur D. Little, Inc. [ADL], 2001, p. 17) to $2,500 per kilowatt (U.S. Department of Energy [DOE], 1998, as cited in Lipman, Delucchi, and Friedman, 2000, p. 18). This translates into $30,000 to $150,000 for the 60-kW systems that will be needed for standard-size vehicles. However, many analysts project that fuel-cell costs will drop dramatically because of design and manufacturing improvements and volume production. For example, Ford Motor Company and Directed Technologies, Inc. (DTI) detail a product specification and manufacturing process that result in costs of $33 to $43 per kilowatt (fuel-cell stack and auxiliaries combined) at high-production volumes (Oei et al., 2000). The following paragraphs describe the key studies; Table 4.9 summarizes their findings.

Arthur D. Little (ADL) 2001 Study. This study employs the Delphi method to solicit opinions from experts on the cost and performance of PEM fuel-cell vehicles over the next 30 years. It provides costs for a fuel-cell system that could use hydrogen generated from gasoline by an on-board fuel reformer. Such fuel cells require higher loadings of precious metals because of impurities that remain after the reforming process. The report concludes that future costs for a complete system (stack and auxiliaries) could drop to $111 per kilowatt. The key cost driver with

Table 4.9

Summary of Studies on Fuel-Cell Costs

Study	Sponsor	Platinum Loading (mg/cm^2)	High-Volume Cost
Arthur D. Little, 2001	California Energy Commission	0.8[a]	$111/kW for stack plus auxiliaries
Fuel Cell Technical Advisory Panel, 1998	CARB	0.25	$20/kW for stack; $15/kW for auxiliaries
Oei et al. (Ford/DTI), 2000	Ford and U.S. DOE	0.25	$19-$28/kW for stack
Lipman, Delucchi, and Friedman, 2000	Kirsch Foundation and Union of Concerned Scientists	0.25	$39-$43/kW for stack plus auxiliaries

[a]The ADL design requires rhodium as well as platinum.

regard to fuel cells is the amount of precious materials required. As shown in Table 4.9, the ADL study assumes 0.8 grams of platinum and rhodium per square centimeter of active fuel-cell area.

According to the ADL study (p. 17), experts from Ballard, Ford, and DaimlerChrysler have predicted the cost of the Ballard Mark 900 fuel-cell module to be $50 to $60 per kilowatt in large volumes (approximately 300,000 units per year). This range suggests that the platinum loading in the Ballard design is substantially less than 0.8 gm/cm^2.

Kalhammer et al. (Fuel Cell Technical Advisory Panel, FCTAP) 1998 Study. CARB established the FCTAP in 1996 to assist it in assessing the status and prospect of fuel cells for auto use. A study was completed, and a report was issued in July 1998. The research method was to "acquire information from responses to a widely distributed questionnaire and in visits with organizations considered leaders in key aspects of fuel cell electric engine and vehicle development." The panel concluded:

The Panel's visits and discussions with leading developers and potential manufacturers of PEM fuel cell components and stacks made clear that the fundamental technical barriers to the development of automotive fuel cell stacks have been overcome by the advances achieved over the past 5–7 years. The large increases in the specific performance of PEM cells and stacks also have lowered the cost barriers to the point where future mass production may be able to meet the stringent cost goals for critical cell components and stacks intended for automotive applications (Kalhammer et al., 1998, p. III-21).

The FCTAP went on to predict:

[On] the basis of the performance already achieved with preprototype stacks, there appear to be reasonable prospects for meeting the $20/kW ($1000/50kW) cost target for automotive fuel stacks if production reaches about 100,000 to 200,000 units per year. At this level, the most critical stack components—membrane, MEA and separator plate—will reach production volumes that justify true mass manufacturing methods (Kalhammer et al., 1998, p. III-54).

Oei et al. (Ford/DTI) 2000 Study. This study considers four detailed designs and assesses the cost based on high-volume production (500,000 units per year) with cost reductions gained by applying DFMA (design for manufacturing and assembly) techniques. The DFMA process has been formally adopted by Ford as a systematic means for the design and evaluation of cost-optimized components and systems. It combines historical cost data and manufacturing acumen accumulated by Ford since the earliest days of the company (Oei et al., 2000, p. 3-1).

Ford/DTI project that costs for PEM stacks will range from $19 to $28 per kilowatt in high-volume production. The range is due to the overall stack power rating (e.g., 60 to 80 kW). Costs include material, manufacturing, and assembly costs, as well as markups to reflect profit, scrap, R&D, and administrative costs.

As we discuss in more detail below, the Ford design incorporates a number of well-researched but still untested features. Learning through manufacturing experience could reduce costs further, but Lipman notes that "the estimation methodology used by DTI was specifically designed to identify the lowest cost PEM stack design configuration, and the choice of a production volume of 300,000 units per year suggests that it would be difficult to construe a lower cost case" (Lipman, Delucchi, and Friedman, 2000, p. 19). In Ford and DTI's view, further cost reductions will require important technological advances. They conclude that further cost reductions will require a reduction in platinum catalyst loading or in gas diffusion electrode costs, because the costs of other components are based on "mature" manufacturing technologies (Oei et al., 2000, p. 3-47).

Lipman, Delucchi, and Friedman 2000 Study. To project fuel-cell costs in the near term, Lipman, Delucchi, and Friedman use an experience curve, which is a way to relate unit cost (in this case $/kW) to cumulative production. They start with a cost of $2,000 per kilowatt and 5,000 kW of cumulative production in 2006 (p. 19); costs then decline 22 percent for each doubling of production (this corresponds to a 78 percent experience curve). They note that experience curves typically range from 70 to 90 percent.[10] Costs decline until they reach a lower bound based on DTI's estimates. Lipman, Delucchi, and Friedman reduce cell peak power density roughly 15 percent from that assumed by DTI and arrive at a lower bound of $39 to $43 per kilowatt for the stack and auxiliaries combined.

Figure 4.6 illustrates the experience curve of Lipman, Delucchi, and Friedman for a 25-kW PEM fuel-cell system. As cumulative production increases, costs fall rapidly and approach $60 per kilowatt.

---

[10]Note that these experience curves are for the successful products. For many products, the technology or costs did not allow commercial production that was long enough for an experience curve to be observed.

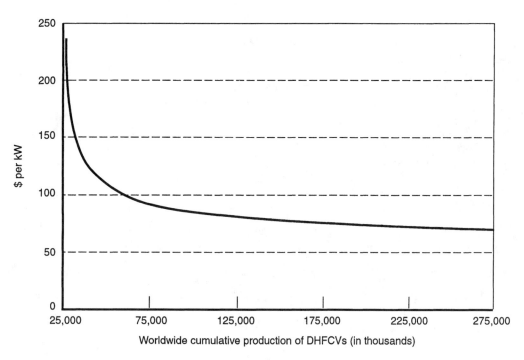

SOURCE: Based on Lipman, Delucchi, and Friedman, 2000.

**Figure 4.6—PEM Fuel-Cell System Costs as a Function of
Accumulated Production Quantity**

Fuel-Cell Components

Membrane Electrode Assembly Cost and Design. The biggest contributor to the cost of a PEM fuel-cell stack is the platinum (which serves as a catalyst). Platinum accounts for a significant portion of the MEA cost in any design. For the design specified in the Ford/DTI study (Oei et al., 2000), the cost of the platinum represents 75 percent of the total cost of the MEA.

Table 4.10 compares the platinum loading in the MEA for the relevant studies. The ADL 2001 study assumes a design with a large amount of platinum ($0.8 \ mg/cm^2$), which easily accounts for half of the MEA cost. All the other studies assume far less for future-year designs. DOE claims a breakthrough of $0.15 \ mg/cm^2$ for experimental MEAs in laboratory quantities from a 3M design.[11] The Ford/DTI study (Oei et al., 2000) assumes $0.25 \ mg/cm^2$, and the FCTAP study (Kalhammer et al., 1998) cites designs that range from 0.25 to $0.35 \ mg/cm^2$. A Los Alamos National Laboratory study (LANL, 2002) describes PEM fuel cells with low platinum loads (0.11 and $0.15 \ mg/cm^2$) that have been tested to 4,000 hours. One reason that the ADL study's loading

_____

[11]In June 2001, DOE announced that 3M was selected for a $7.7 million cost-shared contract to develop MEAs with improved cathodes, high-temperature membranes, and optimized gas diffusion layers. Improved flow fields and MEA fabrication processes will also be developed.

Table 4.10

Platinum Loading in Membrane Electrode Assembly

Study	Platinum Loading (mg/cm^2)	MEA Cost in High Volume ($/kW)	Comment
Arthur D. Little, 2001	0.8	--	Assumes platinum is $13.50/gm
Oei et al. (Ford/DTI), 2000	0.25	10-11	Assumes platinum is $12.86/gm
Los Alamos National Laboratory, 2002	0.11-0.15	--	--
Kalhammer et al. (FCTAP); 1998. Designs reviewed:			
Gore	0.3	--	--
Johnson Matthey	0.35	<15	--
3M	0.25	5-10	Assumes production of >1 million m^2/yr

is much higher is that it is based on a fuel cell that can accept hydrogen reformed from gasoline or methanol. Such fuel cells likely require more platinum and also rhodium to protect against impurities that remain after the reforming process. A DHFCV is less susceptible to this type of catalyst poisoning, so lower catalyst loadings are possible.

These studies suggest that it is highly likely that an MEA with 0.25 mg/cm^2 can be built for automotive use, and we have adopted this catalyst loading in our analysis. Information on the durability of such cells is not publicly available. It is possible that the Ballard fuel-cell system currently being tested by Toyota and GM and the fuel cell in Honda's vehicles contain levels approaching 0.25 mg/cm^2, but such data are not publicly available. Even if the levels in these cells are near 0.25, their real-world durability is still being investigated. We thus cannot be entirely sure that fuel cells with 0.25 mg/cm^2 of platinum will work for automotive use. Designs with even lower loadings, however, suggest that 0.25 mg/cm^2 is not a wildly optimistic assumption.

The low MEA cost in the Ford/DTI study (Oei et al., 2000) is based on a number of other design features that are well documented and researched but have not yet been produced. For example, Ford/DTI propose a composite polymer electrolyte membrane rather than conventional homogeneous membranes (p. 3-5). The projected cost savings are 98 percent of current homogeneous membrane costs. The electrodes in the Ford/DTI study are constructed from an advanced carbon paper that is still under development (p. 3-11). Highly automated manufacturing techniques are used that might not pan out as well as envisioned. Ford/DTI did, however, build some cushion into their estimates. For example, projected costs are inflated by 10 percent to ensure "appropriate conservatism" (p. 3-44), and the lowest projected advanced carbon paper costs were not used (p. 3-14). Ford/DTI believe their projections are based on "accepted manufacturing techniques and realistic technical parameters" (p. 3-19). In their view, the projections would require a minimum of technological advancement (p. 3-46).

Ford/DTI conclude that the MEA for a direct hydrogen fuel-cell stack can be built for $10 to $11 per kilowatt at high-volume production. CARB's FCTAP concludes that costs could run from $5 to $10 per kilowatt.

Bipolar Plate Cost and Design. The role of the bipolar plate (also called the separator plate) is to provide support for the fuel-cell membrane as well as to uniformly distribute the hydrogen on one side of the membrane and the oxygen on the other side. The plate also has a role in conducting heat to allow for proper cooling. It must be corrosion resistant and able to operate under pressure (Dayton, 1999; Bulk Molding Compounds, Inc., 2001).

The FCTAP 1998 study cites a variety of approaches and designs for bipolar or separator plates, as well as uncertainty as to the cost of this component. A number of different types of materials could be used for the bipolar plates, but the four main considerations are

1. corrosion-resistant metals
2. coated metals
3. carbon/graphite
4. compatible polymers.

The Ford/DTI study assumes a metallic bipolar plate made of 316 stainless steel (which contains high percentages of chromium, nickel, and molybdenum). This study considers two types of cell construction: a three-piece design, and a cheaper, unitized (one-piece) separator plate. An earlier ADL study (ADL, 2000) assumes the use of a molded graphite/polymer composite bipolar plate; the plate is molded into two pieces and bonded together.

The FCTAP study does conclude that at least one developer could provide a design that allows the cost of the bipolar plate to be $5 per kilowatt:

> No stack developer appears to have made a final choice, but several approaches look promising, including embossing of impregnated porous carbon, molding of commercially available carbon composites (Energy Partners), embossing of coated metal plates (Allied Signal; Siemens); and perhaps bonding of appropriately shaped metal sheets (H-Power). At present, confident estimates of separator plate costs are still lacking. However, the approaches under development were all selected for their potential to permit low-cost mass manufacturing of plates from inexpensive materials, lending credibility to the $5/kW cost projected by a leading developer (Kalhammer et al., 1998, III-24).

Oak Ridge National Laboratory reports that it is working on a design envisioned to meet a cost goal of $10 per kilowatt. The design being considered is one with thin metallic plates (thinner than a few millimeters) that uses carbon composites instead of graphite, which is more costly to work with (Fuel Cell Catalyst, 2001).

Fuel-Cell Auxiliaries. Auxiliaries are devices that manage air, water, and other thermal issues. In addition, control electronics are required. The cost of the auxiliary equipment supporting a fuel-cell stack has to be considered. Fewer studies project costs for this part of a fuel-cell system. Auto industry goals are $20 per kilowatt. DTI (1998) assesses them to be $14 to $15 per kilowatt; ADL concludes that the balance of plant costs in high-volume production for a gasoline-fueled PEM fuel cell would be $10 per kilowatt of stack power (ADL, 2000).

Summary of Fuel-Cell System Costs. It seems likely that DHFCVs will be available in only very small numbers through 2005 (see Subsection 3.1). We base our fuel-cell system costs between 2006 and 2010 on the manufacturing experience curve developed by Lipman, Delucchi, and Friedman (2000). Our estimates of fuel-cell system costs for this period run from $100 to $150 per kilowatt. For high-production volumes, we use $35 to $60 per kilowatt. The lower bound reflects $20 per kilowatt for the fuel-cell stack and $15 for the auxiliaries. Ford/DTI (Oei et al., 2000) predict stack costs between $18 and $28, and we use $20, which is the low cost for the intermediate-size stacks likely in automotive applications. The upper bound is based on estimates that the Ballard Mark 900 fuel cell will cost $50 to $60 per kilowatt in volume production.

Our review of the literature led us to conclude that the projections that fuel-cell system costs can fall to $35 to $60 per kilowatt in volume production are well grounded. Costs might be even lower if the very low platinum designs currently being investigated pan out. However, it is too early to tell if this will be the case. Manufacturing experience may also reduce costs, although additional reductions may not be large, because most of the nonmaterial costs have been squeezed out by mature manufacturing techniques. However, a number of uncertainties remain. Unexpected technological or production problems may arise, and the durability of automotive fuel cells has not been extensively tested in demanding real-world environments. We use this range for volume production costs in our analysis but note that much remains to be done to demonstrate that these low production costs can actually be realized.

Hydrogen Tank for a Fuel-Cell Vehicle

We considered the cost of the "novel polymer-lined cylinders" described by Thomas et al. (1999). Lipman, Delucchi, and Friedman (2000) estimate their cost to be $510 per kilogram of hydrogen stored at 350 atmospheres (5000 psi) based on current prices for compressed natural gas tanks. Lipman cites a DTI study on high-volume production of such tanks to estimate a lower bound cost of $84 to $163 per kilogram of stored hydrogen. This translates to $336 to $652 for a tank that stores 4 kg. We use this cost range in our estimates of high-volume costs. For the near term, we use the relation between cost and cumulative volume postulated by Lipman, Delucchi, and Friedman (2000).[12] Figure 4.7 shows the resulting relationship for a 4-kg tank.

4.5 PZEV COSTS

CARB reviewed various technological approaches to PZEVs and concluded that the incremental cost relative to a vehicle meeting the SULEV exhaust and near-zero evaporative standards was minor:

> [C]onfidential information from several manufacturers indicates that PZEVs soon to be introduced for sale in California will use a simpler and much less costly combined HC adsorber/catalyst rather than a separate adsorber and an attendant switching valve as had been assumed earlier. Further, the new information indicates that additional catalyst volume will not likely be required as was the case in the first PZEV system certified for sale in California. . . . Manufacturers will still face some increased cost to build increased durability into the emission control components (e.g. increased catalyst loading) in order to avoid excessive repair costs during the 150,000 mile emission warranty period. . . . Based on the information acquired by staff, some additional carbon trap capability will be added, along with improved seals and reconfiguration of some components that do not add large cost. The cost of going to a zero evaporative system from the near zero systems is now estimated to be $10 per vehicle. Taking all of these factors into account, staff now estimates that the necessary hardware modifications to meet PZEV requirements will range from $60 to $85 per vehicle (CARB, 2001d, p. 5).

These estimates are for passenger cars and the smallest category of light-duty trucks (LDT1).[13]

The additional warranty requirements of PZEVs will range between $125 and $150 per vehicle, according to CARB estimates. It is unlikely that all these costs represent incremental costs of PZEVs. To some extent, the warranty causes a transfer of the costs of maintaining the vehicle from consumers to producers and thus does not change the overall resource costs of PZEVs. PZEVs may, however, cost more to maintain over their lifetimes because of the tighter emission standards. The proportion of the warranty that represents a transfer of costs from consumers to producers also depends on the stringency of California's inspection and maintenance (Smog Check) program: A less stringent program requires fewer problems to be repaired in the absence of a warranty, so a larger fraction of the warranty costs are true resource costs. Given these uncertainties, we vary resource costs due to the extended warranty from $0 to $150 per vehicle. This range assumes from none to all of the increased warranty costs represent transfers from consumers to automakers. Thus, our estimate of the incremental cost of a PZEV

---

[12]Tank cost = $321x^{-0.2318}$, where x is cumulative production volume.

[13]Paul Hughes, CARB, personal communication, May 2002.

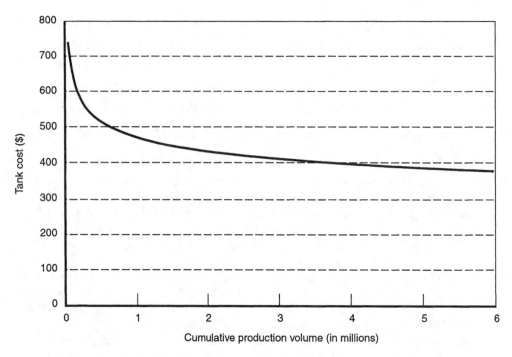

SOURCE: Based on Lipman, Delucchi, and Friedman, 2000.

**Figure 4.7—Cost of Tank That Stores 4 kg of Hydrogen as a Function
of Production Quantity**

relative to a SULEV meeting the near-zero evaporative emission standards ranges from $60 to
$235.

We have not been able to examine in detail the basis for CARB's PZEV cost estimates.
Earlier CARB estimates were higher. In August 2000, CARB put the incremental costs of a
PZEV (relative to a SULEV) at $500 (CARB, 2000b, p. 106). The design improvements outlined
above make a plausible case for the reduced estimates, but further analysis is warranted.

4.6 INCREMENTAL VEHICLE PRODUCTION COSTS

So far we have discussed the costs of individual vehicle components. To calculate the
incremental production costs of vehicles that manufacturers may use to meet ZEV program
requirements, we subtract the costs of components associated with the conventional ICEV and
add in the costs of components associated with each vehicle technology.[14] Indirect costs
associated with each of the components are also dropped and added, and we account for changes
in these costs (e.g., corporate overhead, warranty, and normal profit). Here, we first detail the

---

[14]As an alternative to this approach, one could also build costs of various vehicles from the ground
up and then compare the resulting total vehicle costs. Delucchi et al. (2000) use such an approach.

indirect cost multipliers used in our analysis. We then present estimates of the incremental costs of ZEVs, ATPZEVs, and PZEVs.

Multipliers Used to Calculate Indirect Costs

Analysis of indirect costs is usually framed in terms of the difference between manufacturing cost and final manufacturer's suggested retail price (MSRP). Vyas, Cuenca, and Gaines (1998) break down the components of MSRP and show that the direct manufacturing cost accounts for half of the price (see Table 4.11). A later report (Cuenca, Gaines, and Vyas, 1999) proposes multipliers to translate vehicle cost to vehicle price. Depending on the component, the multipliers range from 1.15 to 2.0: 1.15 to 1.3 for the battery pack, 1.5 for all other components supplied to the auto manufacturer, and 2.0 for components developed solely by the auto manufacturer (e.g., the internal combustion engine). We use these estimates in our analysis. For battery packs and fuel-cell systems, 1.15 and 1.30 are used in calculations of the lower and upper bounds, respectively, of the cost ranges projected for each production volume scenario.[15]

Incremental Vehicle Production Costs

Estimates of the incremental production costs of the various low-emission vehicles and ZEVs are presented in Table 4.12. Costs are incremental to a standard ICEV that meets the SULEV exhaust standard and the near-zero evaporative emission standard. The second column presents estimates of the range into which incremental production costs are likely to fall on average between 2003 and 2007 if annual production volumes are at the low end of the ranges specified in Subsection 3.2. The third column presents the range if annual production volumes are at the high end of the ranges specified in Subsection 3.2. The last column reports incremental costs in volume production. Breakdowns of volume production costs for each vehicle technology are presented in Appendix C, as is an analysis of the sensitivity of the incremental cost of full-function EVs to changes in peak motor power and battery pack size.

As discussed above, the incremental costs of a PZEV are modest. GHEVs will likely cost several thousand dollars more to produce during the first five years of the program. Costs come down in high-volume production but are still expected to exceed standard ICEV costs by $500 to $900 per vehicle. BPEVs will be much more expensive to produce during the initial years of the program. Incremental production costs including the cost of the first battery pack for a full-function EV using NiMH batteries range from $16,700 to $26,100. Incremental costs for PbA

---

[15]CARB's first estimate of PZEV costs included a 10 percent adjustment for indirect costs (CARB, 2000b, p. 106). We assume that CARB's 2001 estimate also includes indirect costs, but the indirect cost multiplier is unknown.

Table 4.11

Breakdown of Manufacturer's Suggested Retail Price for an ICEV

	Share of MSRP (%)	
Cost Category/Subcategory	By Cost Category	By Subcategory
Production	67.0	
Manufacturing		50.0
Warranty		5.0
R&D and engineering		6.5
Depreciation/amortization		5.5
Selling	23.5	
Distribution		20.0
Advertising and dealer support		3.5
Administration and profit	9.5	
Corporate overhead		5.0
Retirement and health benefits		2.0
Profit		2.5
Total	100.0	100.0

SOURCE: Vyas, Cuenca, and Gaines, 1998.

Table 4.12

Incremental Vehicle Production Costs (relative to an ICEV meeting SULEV exhaust and near-zero evaporative emission standards) (estimated ranges in brackets; $1000s per vehicle)

	Near-Term Costs		
Vehicle Technology	Low-Volume Scenario	High-Volume Scenario	High-Volume Production
	2003-2007	**2003-2007**	
PZEV	[0.06, 0.23]	[0.06, 0.23]	[0.06, 0.23]
GHEV	[2.4, 3.3]	[2.3, 2.9]	[0.5, 0.9]
Full-function EV			
NiMH	[22.5, 26.1]	[16.7, 19.8]	[5.1, 9.9]
PbA	[12.7, 15.2]	[8.0, 9.6]	[1.8, 4.3]
City EV			
NiMH	[12.0, 13.4]	[8.5, 9.7]	[1.7, 3.4]
PbA	[8.6, 9.9]	[5.3, 6.0]	[0.7, 1.5]
	2006-2010	**2006-2010**	
DHFCV	[11.4, 15.2]	[8.3, 11.0]	[1.2, 3.9]

batteries and for city EVs are lower but remain very large between 2003 and 2007. Even in volume production, production costs remain substantially above those for ICEVs. The lowest incremental cost in high-volume production is $700 to $1,500 for city EVs with PbA batteries. But these numbers should be used with caution because, as discussed above, there are drawbacks to designing vehicles with PbA batteries. DHFCVs will also likely be much more expensive when they are first introduced. Because of low production volumes and the developing technology, we estimate that the incremental costs of DHFCVs will run from $8,300 to $15,200 between 2006 and 2010. But in the long run, DHFCV technology looks more promising than BPEV technology does. Projections based on what is currently known about fuel cells show that incremental production costs may come down to $1,200. However, we cannot rule out the possibility that production costs will remain nearly $4,000 above those of an ICEV.

4.7 INCREMENTAL LIFECYCLE COSTS

So far we have discussed the incremental production cost of advanced technology vehicles, but their overall social cost also depends on their incremental lifetime operating costs, by which we mean the cost of operating and maintaining the vehicle relative to the cost of operating and maintaining a baseline ICEV. We add incremental lifetime operating costs to incremental production costs to estimate incremental lifecycle cost.

We begin by presenting and justifying the parameters we use to estimate the incremental lifecycle costs, starting with full-function EVs and city EVs and then moving to DHFCVs and GHEVs. Any incremental lifetime operating costs of PZEVs (due, for example, to increased warranty costs) are captured in the production costs estimated above. Thus, the PZEV incremental lifecycle cost is only the incremental production cost. After presenting the parameters, we discuss the method we use to predict incremental lifecycle costs using these parameters. Finally, we present the results of the analysis.

Parameters Used to Estimate Incremental Lifecycle Costs of Full-Function and City EVs

The incremental lifecycle costs of BPEVs are determined by

- incremental initial vehicle costs (including first battery and off-board charger)
- vehicle life
- replacement costs and salvage value of batteries
- EV and ICEV fuel costs
- EV and ICEV repair and maintenance costs
- charger installation costs
- the discount rate.

Table 4.13 lists the parameters that determine the incremental costs for full-function EVs in each of these areas, along with plausible ranges for their values between 2003 and 2007 and when these vehicles are in high-volume production. Parameters for full-function EVs using both NiMH and PbA batteries are shown. Table 4.14 lists parameter values for city EVs with both types of batteries. We now discuss the rationale for each parameter range.

Incremental Vehicle Cost. The incremental vehicle costs between 2003 and 2007 and in volume production are taken from Table 4.12. For both the full-function and the city EVs, they include the cost of the first battery pack.

Vehicle Life. Using data from the *Transportation Energy Data Book* and other sources, Delucchi et al. (2000, p. 55) estimate an average vehicle life of 150,000 miles. In its analysis of

Table 4.13

Parameters Used to Calculate Lifetime Operating Cost of a Full-Function EV Relative to a Comparable ICEV[a]

Parameter	Full-Function EV with NiMH Batteries		Full-Function EV with PbA Batteries	
	2003-2007	Volume Production	2003-2007	Volume Production
Initial Vehicle Cost ($) (including first battery and off-board charger)	16,700-26,100	5,100-9,900	8,000-15,200	1,800-4,300
Vehicle Life				
Lifetime vehicle miles traveled	150,000	150,000	150,000	150,000
Vehicle life in years	15	15	15	15
Parameters Affecting Battery Replacement Cost				
Pack size (kWh)	30	24	30	24
Battery cost ($/kWh)	305-360	233-258	115-154	114
Cycle life (1000 cycles)	1.0-1.5	1.5	0.6-1.0	1.0
Range (miles/cycle)	90-110	90-110	90-110	90-110
Cost surcharge for installing subsequent packs ($/installation)	500	500	500	500
Salvage value ($/kWh)	0-40	0-40	-15 to 5	-15 to 5
Parameters Affecting Fuel Costs				
EV efficiency (mi/kWh AC)	2-3	2.5-3.5	2-3	2.5-3.5
ICEV efficiency (mi/gal)	25-30	25-30	25-30	25-30
Electricity price ($/kWh)	0.035-0.05	0.07-0.09	0.035-0.05	0.07-0.09
Gasoline price ($/gal)	0.56-1.49	0.56-1.49	0.56-1.49	0.56-1.49
Other Costs				
EV repair and maintenance (% of ICEV costs)	65-70	65-70	65-70	65-70
ICEV maintenance ($/mile)	0.06	0.06	0.06	0.06
Charger installation ($)	500-1,500	500	500-1,500	500
Discount Rate (%)	4	4	4	4

[a]The comparable ICEV is mid-sized, meets the SULEV and near-zero evaporative emission standards, and averages 25 to 30 miles per gallon.

lifecycle costs, CARB assumes that vehicles travel 117,000 miles over their lifetimes (CARB, 2000b, p. 98), which is close to CARB's definition of a "useful vehicle" life, rather than actual miles driven before a vehicle is retired. Thus, for a full-function EV, we set lifetime miles traveled at 150,000 miles. CARB also assumes a vehicle life of 10 years; we set vehicle life to 15 years to correspond to the higher lifetime miles traveled that we assumed in our analysis.

Because of the limited top speed and range of city EVs, they are unlikely to be driven as far on average as full-function EVs over their lifetimes. Based on the National Personal Transportation Survey, Reuyl and Schuurmans (1996) estimate that vehicles with a daily range of 50 to 60 miles can be used for 50 percent of the miles driven by the average driver. We thus assume that city EVs will be driven 75,000 miles over their lifetimes; CARB assumes 88,000 (2000b, p. 113).

Table 4.14

Parameters Used to Calculate Lifetime Operating Cost of a City EV
Relative to a Comparable ICEV[a]

Parameter	City EV with NiMH Batteries		City EV with PbA Batteries	
	2003-2007	Volume Production	2003-2007	Volume Production
Initial Vehicle Cost ($) (including first battery and off-board charger)	8,500-13,400	1,700–3,400	5,300-9,900	700-1,500
Vehicle Life				
Lifetime vehicle miles traveled	75,000	75,000	75,000	75,000
Vehicle life in years	15	15	15	15
Parameters Affecting Battery Cost				
Pack size (kWh)	10	8	10	8
Battery cost ($/kWh)	305-360	233-258	115-154	114
Cycle life (1,000 cycles)	1.0-1.5	1.5	0.6-1.0	1.0
Range (miles/cycle)	50-60	50-60	50-60	50-60
Cost surcharge for installing subsequent packs ($/installation)	250	250	250	250
Salvage value ($/kWh)	0 to 40	0 to 40	-15 to 5	-15 to 5
Parameters Affecting Fuel Costs				
EV efficiency (mi/kWh AC)	4-5	5-6	4-5	5-6
ICEV efficiency (mi/gal)	40	40	40	40
Electricity price ($/kWh)	0.035-0.05	0.07-0.09	0.035-0.05	0.07-0.09
Gasoline price ($/gal)	0.56-1.49	0.56-1.49	0.56-1.49	0.56-1.49
Other Costs				
EV repair and maintenance (% of ICEV costs)	65-70	65-70	65-70	65-70
ICEV maintenance ($/mile)	0.045	0.045	0.045	0.045
Charger installation ($)	0	0	0	0
Discount Rate (%)	4	4	4	4

[a]The comparable ICEV is small, meets the SULEV and near-zero evaporative emission standards, and averages 40 miles per gallon.

Parameters That Affect Battery Replacement Cost. The cost of the batteries (excluding the first battery pack) needed over a vehicle's lifetime is determined by

- the battery pack size
- the cost of the batteries in dollars per kilowatt-hour
- the cycle life of the battery
- the miles driven on a battery cycle
- the cost markup for buying and installing subsequent battery packs
- the salvage value of batteries.

We discuss each in turn.

As described in Subsection 4.1, we set the 2003-through-2007 battery pack size to 30 kWh for a full-function EV and to 10 kWh for a city EV. It seems plausible that the efficiency of EVs (in terms of kilowatt-hours per mile) will increase in the future (see discussion below). We increase efficiency by 20 percent for our estimates of lifecycle costs in volume production so that the battery pack can be smaller and maintain the same range. Accordingly, we reduce the full-function EV and city EV battery packs to 24 and 8 kWh, respectively, for the high-production volume cost calculations.

The above analysis of battery costs led us to conclude that NiMH battery packs (including battery auxiliaries and indirect costs) would cost from $425 to $460 per kilowatt-hour between 2003 and 2007. The range for PbA batteries is $130 to $170. The cost of replacement batteries for vehicles sold between 2003 and 2007 may well be lower than it was when the vehicle was built due to battery production volumes increasing over time. We assume that by the time the battery is replaced, 50 percent of the difference between the 2003-2007 costs and the midpoint of the high-volume cost range will have been eliminated. The result: NiMH replacement costs will be $305 to $360, and PbA replacement costs, $115 to $154. In the volume production scenarios, we allow NiMH battery pack costs to range from $233 to $258 per kilowatt-hour and set PbA battery costs at $114 per kilowatt-hour.

The EV's range (90 to 110 miles for a full-function EV and 50 to 60 miles for a city EV) multiplied by the battery cycle life is a starting point for the miles driven over one pack's lifetime, which in turn determines the number of packs needed over a vehicle's lifetime. We reduce the resulting estimate of the miles driven using a single pack by 20 percent, because vehicle range is determined by taking the battery down to empty, whereas cycle life is usually determined by taking the battery down to a 20 percent charge.

We put NiMH cycle life at 1,000 to 1,500 cycles between 2003 and 2007 and at 1,500 cycles in the high-volume cost scenario. CARB's Battery Technology Advisory Panel (BTAP) report (Anderman, Kalhammer, and MacArthur, 2000, p. 37) cites one auto manufacturer's projection of 1,200 cycles and 100,000 miles at 77°F and 1,100 cycles at 95°F from bench tests that have been partially corroborated by field tests of battery packs in vehicles. While cycle life deteriorates when the pack is continuously exposed to extreme heat (113°F), the important implication is that a NiMH battery pack that is properly thermally managed can now achieve 100,000 miles. In reference to a Panasonic NiMH battery,[16] the BTAP report optimistically forecasts that "it seems quite possible that 1,000 to 2,000 cycles at 100 percent depth of discharge

---

[16]At the November 1999 UC Davis Conference on EVs, Panasonic EV reported that the RAV4-EV (NiMH) had already achieved 100,000 kilometers of driving on one pack in tests in Japan. Furthermore, Panasonic EV suggested that longer life can be expected with "economy charging," i.e., intermittent full charging in order to reduce deterioration of the NiMH battery electrode.

can eventually be achieved, depending on the battery's initial power versus the car's requirements" (Anderman, Kalhammer, and MacArthur, 2000, p. 31).

PbA batteries currently achieve roughly 600 cycles, but Panasonic has been developing designs with longer cycle lives, and 1,000 cycles appears to be possible (Matsushita Electric, 2000).[17] We set PbA cycle life to 600 to 1,000 cycles between 2003 and 2007, and to 1,000 cycles in the high-volume scenarios.

Following CARB (2000b, p. 100), we set the cost of handling and installing replacement battery packs at $500 per pack.

Batteries may have salvage value at the end of their lives. Battery experts at Southern California Edison told Delucchi et al. (2000, p. 45) that used NiMH batteries would be at least as good for load-leveling, backup, and other utility applications as the new PbA packs they currently buy for such uses. Thus, the salvage value for NiMH batteries might be as high as $100 per kilowatt-hour (roughly the current cost of new PbA batteries), at least until the supply of retired NiMH batteries saturates the market. The secondary market for NiMH batteries is quite uncertain, however. CARB (2000b, p. 100) assumes that the salvage value is $40 per kilowatt-hour for NiMH batteries, but Delucchi et al. (2000) believe the batteries will lose their remaining value very quickly and that there thus will be little market for them. They believe the batteries will have to be recycled at a small net cost at low volumes and will have a small net salvage value at medium volumes. In our analysis, we vary the salvage value of NiMH batteries from $0 to $40 per kilowatt-hour.

CARB (2000b, p. 100) believes a secondary market for used PbA batteries at meaningful volumes is speculative and assumes a salvage value of $3 per kilowatt-hour—the value of the materials in the battery. Delucchi et al. (2000, p. 46) assumes that PbA batteries will have negative salvage value—that is, they will cost between $5 and $15 to recycle. In our analysis, we vary PbA salvage values from -$15 to $5 per kilowatt-hour.

The last battery pack installed in the vehicle might not be fully used at the end of the vehicle life. Following Delucchi et al. (2000, p. 69), we set the salvage value of the unused portion of the pack equal to 70 percent of the cost of batteries to the manufacturer.

Parameters That Affect Fuel Costs. The cost of electricity, the cost of gasoline, the wall-to-wheels (or AC) efficiency of the EV, and the fuel efficiency of the ICEV determine the lifetime fuel costs of a full-function EV relative to an ICEV.

---

[17]"The newly developed EV battery lasts approximately twice as long and offers about twice as much power as existing lead-acid EV batteries. It can be discharged and recharged approximately 1,000 times, which is about twice as many cycles as the company's existing lead-acid EV battery products" (Matsushita Electric, 2000).

The number of BPEVs is relatively small between 2003 and 2007, and we assume that BPEVs can be charged off-peak using existing power generation, transmission, and distribution facilities. As shown in Appendix D, the costs of power in such a situation might be expected to run from $0.035 to $0.05 per kilowatt-hour. If EVs become a significant part of the transportation system, they will use a significant amount of power, requiring additional generation, transmission, and distribution facilities.[18] For volume production, we thus put electricity costs at $0.07 to $0.09 per kilowatt-hour, which covers the capital cost of the power plant and the transmission and distribution system (see Appendix D). Because we are interested in resource costs, we do not include taxes or charges imposed to recover costs during California's recent energy crisis. Likewise, CARB does not include taxes in its analysis of operating costs. CARB assumes an electricity cost of $0.05 per kilowatt-hour; however, it also presents a scenario in which costs are $0.075 per kilowatt-hour (CARB, 2000b, p. 107).

There is a good deal of uncertainty regarding gasoline prices over the lives of the vehicles. We assume prices at the pump will vary between $1.00 and $2.00 per gallon, which is consistent with prices observed in recent years in California. Federal and state fuel excise taxes currently total $0.363 per gallon, and sales taxes on gasoline average roughly 8 percent in the state (CARB, 2000b, p. 107). We thus assume that gasoline costs vary from $0.56 to $1.49 per gallon, net of taxes.

EV efficiency (in miles per kilowatt-hour AC) is based on data developed by Southern California Edison for the EVs produced by the large manufacturers.[19] Our figures for ICEV fuel efficiency reflect a rough average of the fuel efficiencies of ICEVs that are comparable in size to full-function EVs and city EVs (27.5 miles per gallon for mid-size ICEVs and 40 miles per gallon for very small ICEVs).

The efficiency of EVs may improve over future design cycles and thus might be higher in volume production than it will be between 2003 and 2007. CARB assumes that efficiency increases from about 2.2 miles per kilowatt-hour in 2003 to roughly 4 miles per kilowatt-hour in volume production for a four-passenger full-function EV (CARB, 2000b, p. 111). This improvement is driven by a substantial reduction in aerodynamic drag, lower-loss tires, a higher-efficiency drive system, and substantial improvements in charging efficiency. While efficiency may well improve, we think it likely that CARB is overly optimistic. For example, vehicles with very low aerodynamic drag are technologically feasible but are not assured of success in the

---

[18]For example, to charge 100,000 full-function EVs overnight will require a 250-megawatt power plant—a good-sized plant. (If each full-function EV requires 30 kWh per night, 100,000 full-function EVs require 3,000 megawatt-hours. And 3,000 MWh/12 hours = 250 megawatts.)

[19]We determined the range using the efficiencies remaining after the vehicles with the highest efficiency (the EV1) and lowest efficiency (the EPIC) were discarded (see http://ev.inel.gov/fop).

marketplace. We increase EV efficiency roughly 20 percent from the levels used to estimate costs between 2003 and 2007, yielding a battery pack that is 20 percent smaller but allows the same range.

EV Repair and Maintenance Costs. It is often argued that EVs will have lower repair and maintenance costs than ICEVs do, because EVs have fewer components and fewer moving parts. While this may be so in the long term, we follow Dixon and Garber (1996, p. 244) in assuming there are several reasons to be skeptical in the near term:

- EV technology is evolving rapidly, but projections of EV repair and maintenance costs may be based implicitly on costs for stable designs. Components will probably be less reliable in the early years of commercial scale production than in later years.
- Several manufacturers have had problems with their battery packs and have had to replace battery modules, an expensive process. Installing replacement packs may also be time consuming and costly.
- Repair and maintenance costs of ICEVs appear to be declining.

Delucchi et al. (2000, p. 245) put EV repair and maintenance costs at approximately 70 percent of ICEV costs; CARB (2000b, p. 108) uses 66 percent in its analysis. We let the proportion vary from 65 to 70 percent. The ICEV repair and maintenance cost we use for reference for the full-function EV is $0.06 per mile (as in Delucchi et al., 2000, p. 245, and CARB, 2000b, p. 108). We reduce the repair and maintenance cost by 25 percent for the smaller ICEV used in the city EV incremental-cost calculation.

Charger Installation. We assume that the off-board charging equipment is included in the cost of the EV, but that the off-board charger must be installed at the business or residence where the vehicle is charged. Installation often requires significant extension or modification of the business/residence electrical system and may require structural modifications. There appears to be a great deal of variation across customers in charger installation costs. Nissan reports an average cost (based on a fairly small number of installations) of $3,000 for installation of a full-function EV charger inside the garage of a single-family home (Atkins et al., 1999); CARB sets the cost of the charger plus installation at $1,500 in 2003 and $750 in volume production. Our analysis assumes installation costs range from $500 to $1,500 between 2003 and 2007. Average installation costs in volume production should be lower—in part because some households and businesses will already have installed the 220-volt line used by most full-function EV chargers. We set installation costs at $500 for volume production.

It appears that city EVs will likely be offered with a number of different charging options, ranging from simply plugging the vehicle into a standard 110-volt outlet to using a rapid charging system that requires the same 220-volt, 30-ampere outlets used by full-function EVs. The 110-volt option requires no charger installation costs, but it may take up to 8 hours to charge the 10-kWh battery. The rapid-charge option will charge the battery very quickly but may require installation costs as high as those for full-function EVs. We set charger installation costs to zero in our analysis. The key requirement for any EV is that it be able to charge overnight, and the city EV will be able to do so using a standard 110-volt circuit.

   Discount Rate. The discount rate translates costs incurred in the future to present-day costs. Because we are interested in the resource costs of advanced technology vehicles to society as a whole, we set the discount rate to the social (as opposed to the private) discount rate.[20] The social discount rate is defined as the social opportunity cost of capital and is often based on the long-run, risk-free (or low-risk) return on capital (Tietenberg, 1992, p. 88). Examples include the return on U.S. Treasury bonds. Estimates of the long-run, risk-free return are typically 3 to 5 percent, net of inflation.

Parameters Used to Estimate Incremental Lifecycle Costs of DHFCVs

   Table 4.15 lists the parameters used to estimate the incremental lifecycle costs of a DHFCV relative to an ICEV that meets the SULEV exhaust and near-zero evaporative emission exhaust standards. The key parameters in the incremental operating costs are the DHFCV's fuel economy in miles per kilogram of hydrogen and the cost of hydrogen in dollars per kilogram. The Methanol Institute summarizes the findings of eight studies on the expected fuel economy of DHFCVs.[21] The efficiencies range from 55 to 106 miles per equivalent gallon of gasoline, and the Methanol Institute uses 65 to 85 miles in its analysis. The two measures—miles per equivalent gallon of gasoline and miles per kilogram of hydrogen—are very close, and we use 65 to 85 miles per kilogram for our cost projections for 2006 to 2010. For high-volume lifecycle cost projections, we set the efficiency at 85 miles per kilogram.

   Appendix D derives an estimate of the cost of hydrogen fuel. We assume that hydrogen is sold at filling stations that generate hydrogen from natural gas piped to the site. Assumptions on the cost of natural gas and the cost of the reformer that generates hydrogen from natural gas result in hydrogen costs of $2.32 to $3.16 per kilogram between 2006 and 2010. The costs fall to $1.85 to $2.33 per kilogram for high-volume production of reformers. To put these numbers in perspective, $2.00 per kilogram corresponds to $0.027 per mile in a vehicle that gets 75 miles per kilogram. Fuel cost for a mid-size ICEV that

---

[20]Consumers are thought to apply much higher discount rates when buying durable goods. See, for example, Kavalec, 1996, p. 30.

[21]See http://www.methanol.org/fuelcell/special/contadini_pg1.html, Tables 1 and 2.

Table 4.15

**Parameters Used to Calculate Lifetime Operating Costs of a DHFCV
Relative to a Comparable ICEV[a]**

Parameter	2006-2010	Volume Production
Initial vehicle cost ($)	8,300-15,200	1,200 - 3,900
Vehicle efficiency		
DHFCV (mi/kg of hydrogen)	65-85	85
ICEV (mi/gal)	25-30	25-30
Fuel cost		
Gasoline ($/gal)	0.56-1.49	0.56-1.49
Hydrogen ($/kg)	2.32-3.16	1.85-2.33
Maintenance costs ($/mi)		
DHFCV	0.06	0.06
ICEV	0.06	0.06

[a]The comparable ICEV used in the calculations is mid-size, meets the SULEV and near-zero evaporative emission standards, and averages 25 to 30 miles per gallon.

averages 27.5 miles per gallon when gasoline (net of taxes) is $1.00 per gallon is $0.036 per mile.

Little information is available on the maintenance costs of DHFCVs relative to ICEVs. A DHFCV's electric power train has fewer parts than an ICEV's drivetrain and may be cheaper to maintain. The reliability and maintenance costs of a PEM fuel cell, however, are uncertain. Lacking better information, we set the maintenance costs of DHFCVs equal to those for ICEVs.

Parameters Used to Estimate Incremental Lifecycle Costs of GHEVs

GHEVs may be cheaper to operate than standard ICEVs because of their higher gas mileage. Following CARB (2000b, p. 114), we set the efficiency of a four-passenger GHEV to 45 miles per gallon between 2003 and 2007 and assume that improvements in vehicle design will increase the fuel economy to 55 miles per gallon in volume production (see Table 4.16).[22] These fuel economies are 50 and 57 percent greater, respectively, than the 30 and 35 miles per gallon that CARB (2000b, p. 115) assumes for a comparable ICEV.

In its initial analysis of hybrid costs, CARB concluded that hybrid maintenance costs may be higher than those of standard gasoline vehicles because hybrids contain both a conventional and an electric drive system. CARB noted that little was known about hybrid operating costs, however, and "in the absence of more specific information," set maintenance costs at $0.075 per mile (CARB, 2000b, p. 108). CARB later reexamined the maintenance costs of hybrids and

---

[22]CARB's estimate of 45 miles per gallon is based on published fuel mileage figures for currently available hybrids.

Table 4.16

Parameters Used to Calculate Incremental Lifecycle Cost of a GHEV Relative to a Comparable ICEV[a]

Parameter	2003-2007	High-Volume Production
Initial vehicle cost ($) (including first battery and off-board charger)	2,300-3,300	500-900
Vehicle efficiency (mi/gal)		
GHEV(ATPZEV)	45	55
ICEV	30	35
Fuel costs ($/gal)	0.56-1.49	0.56-1.49
Maintenance costs ($/mi)		
GHEV(ATPZEV)	0.06-.078	0.06-.065
ICEV	0.06	0.06

[a]The comparable ICEV used in the calculations is mid-size, meets the SULEV and near-zero evaporative standards, and averages 30 miles per gallon between 2003 and 2007 and 35 miles per gallon when GHEVs are in volume production.

concluded that incremental maintenance costs would be zero. CARB noted that there may be some incremental maintenance costs associated with the battery or electric drive components (e.g., the 2002 Toyota Prius requires a coolant change for the electric power inverter at 34,000 miles) but that there may be less brake wear due to regenerative braking (CARB, 2001h, p. 167).

CARB never explicitly addresses the costs associated with maintaining and possibly replacing the battery in hybrids. There is uncertainty over whether one battery will last the entire life of the vehicle. As noted in Subsection 4.3, a recent report put the life expectancy of a NiMH hybrid battery at five to eight years. Ford reports that the battery in its Escape hybrid will have to be replaced at least once during a 15-year vehicle life. Honda implies that 15 years in service is beyond today's battery technology. Toyota believes that scheduled replacement of the battery will not be required over a 15-year lifetime but that unscheduled maintenance is possible (CARB, 2001h, p. 129).

This uncertainty drives us to develop a range for hybrid maintenance costs. The lower end of the range is $0.06 per mile, equal to the maintenance costs of standard ICEVs. The upper end allows for the accumulation of enough money over the life of the vehicle to replace the battery once. For 2003 to 2007, we take the upper estimate for the cost of the battery ($2,000—see Subsection 4.3), add 30 percent for indirect costs (see Subsection 4.6), and add $125 for installation (lower than the $500 assumed cost to replace the much larger full-function EV pack). The sum translates into $0.018 per mile, resulting in a maintenance cost range that runs from $0.06 to $0.07 per mile. In high-volume production, the pack costs $500, resulting in a range of $0.06 to $0.065 per mile.

Method for Calculating Incremental Lifecycle Costs

We assume that each of the parameters is statistically independent and uniformly distributed on its specified range. We then randomly draw a value from each parameter range and use the resulting set of values to calculate the incremental lifetime operating cost and incremental lifecycle cost of each advanced technology vehicle relative to an ICEV. We repeat this process 10,000 times and report the 5th and 95th percentiles of the distribution. Given assumptions on the distributions of the parameters, if we repeated the process for constructing the interval over and over, this interval would contain the true incremental cost 90 percent of the time. We chose to repeat the process 10,000 times because the distribution is stable to two significant digits with this number.

Findings on Incremental Lifecycle Costs

Results of our analysis of incremental lifetime operating and lifecycle costs are reported in Table 4.17. The second and third columns show our findings on the incremental lifetime operating cost in the near term and in volume production, respectively. The fourth and fifth columns add in the incremental initial cost of the vehicle and show the incremental lifecycle cost. The brackets in each cell contain the 90 percent probability interval for the estimate.

Full-Function EVs. During the first five years of the program, the lifetime operating cost of full-function EVs with NiMH batteries may be higher or lower than that of an ICEV meeting the SULEV exhaust and near-zero evaporative emission standards: As can be seen in the second column of Table 4.17, the 90 percent probability interval runs from -$1,600 to $4,000. Once the high incremental production cost for these vehicles is included, the incremental lifecycle cost ranges from $17,300 to $27,800. Our analysis suggests that even in volume production, the lifecycle cost of full-function EVs with NiMH batteries will remain substantially above that for ICEVs: The incremental cost estimates range from $3,100 to $8,700.

PbA batteries are cheaper than NiMH batteries, but their short cycle life means that the lifetime operating cost will likely be somewhat greater for full-function EVs with PbA batteries than for their NiMH-powered counterparts. Incremental lifecycle cost is a good deal lower, however, because the initial vehicle is cheaper. Between 2003 and 2007, there will still be a substantial lifecycle cost premium for full-function EVs powered with PbA batteries ($9,000 to $18,400). Incremental lifecycle cost may be small in volume production ($200), but there is an equally good chance that incremental lifecycle cost will remain substantial (the upper end of the 90 percent probability interval is $5,100). Thus, even with PbA batteries, the lifecycle cost of full-function EVs is likely to exceed that of ICEVs. What is more, questions remain about whether PbA batteries are feasible for full-function EVs that have a range of roughly 100 miles,

Table 4.17

Incremental Lifetime Operating and Lifecycle Costs of Advanced Technology Vehicles
(incremental over a vehicle meeting SULEV and near-zero evaporative emission standards)
($1000s per vehicle; 90 percent probability intervals in brackets)[a]

Vehicle Technology	Incremental Lifetime Operating Cost		Incremental Lifecycle Cost	
	Near Term	High-Volume Production	Near Term	High-Volume Production
	2003-2007		**2003-2007**	
Full-function EV				
NiMH	[-1.6, 4.0]	[-3.1, 1.0]	[17.3, 27.8]	[3.1, 8.7]
PbA	[-1.1, 5.2]	[-2.5, 5.1]	[9.0, 18.4]	[0.2, 5.1]
City EV				
NiMH	[-1.3, 0.5]	[-1.6, -0.3]	[8.1, 12.9]	[0.5, 2.4]
PbA	[-1.1, 1.0]	[-1.3, -0.1]	[5.1, 9.9]	[-0.4, 1.2]
GHEV	[-1.4, 0.8]	[-1.5, 0.3]	[1.3, 3.7]	[-0.9, 0.4]
PZEV	0	0	[0.06, 0.23]	[0.06, 0.23]
	2006-2010		**2006-2010**	
DHFCV	[-2.1, 2.0]	[-3.4, 0.3]	[7.8, 15.6]	[-1.3, 3.4]

[a]Discounted to the time of vehicle sale using a 4 percent discount rate.

since the size and weight of these batteries make it difficult to design such a vehicle with acceptable space and handling.

City EVs. Mainly because of their smaller battery packs, city EVs have a much lower incremental lifecycle cost than full-function EVs do. Between 2003 and 2007, the operating cost of a city EV with either PbA or NiMH batteries may be above or below that of a small ICEV. Substantial incremental vehicle production cost between 2003 and 2007 still means, however, that incremental lifecycle cost for a city EV remains sizable with either type of battery.

For high-volume production, our analysis predicts that if battery cycle life reaches the most optimistic expectations and vehicle efficiencies improve somewhat, the lifecycle cost of a city EV with PbA batteries could be comparable to that of a small ICEV. As shown in Table 4.17, the 90 percent probability interval ranges from -$400 to $1,200. Again, however, note that manufacturers are choosing to use NiMH batteries and NiCd batteries in their city EVs despite the higher costs, which suggests that PbA batteries do not work well in a city EV design.

GHEVs. The cost of servicing and possibly replacing the battery in a GHEV more than offsets the savings due to this vehicle's higher gas mileage. Our estimates of incremental lifetime operating cost range from -$1,400 to $800 per vehicle for vehicles produced between 2003 and 2007 and from -$1,500 to $300 in high-volume production. Once the incremental cost of the vehicle is added in, the hybrid's lifecycle cost is $1,300 to $3,700 greater than that of a standard ICEV between 2003 and 2007. Incremental lifecycle cost runs from -$900 to $400 in volume production.

PZEVs. The greater incremental lifecycle cost of a PZEV relative to an ICEV that meets the SULEV exhaust and near-zero evaporative emission standards is due only to the increased initial cost of a PZEV. Our estimates range from $60 to $235 from 2003 on.

DHFCVs. Given our assumptions on the cost of hydrogen fuel, fuel efficiency, and vehicle maintenance, the lifetime operating cost of a DHFCV may be several thousand dollars less than that of an ICEV. The midpoint of the estimate for incremental operating cost between 2006 and 2010 is near zero, although there is enough uncertainty in the parameters that the upper end of the 90 percent probability interval reaches $2,000. Very high initial vehicle cost means that the incremental lifecycle cost of DHFCVs remains high through 2010. Plausible improvements in fuel-cell technology and reductions in the cost of providing hydrogen fuel may ultimately reduce the lifecycle cost for this vehicle to below that for an ICEV (the lower end of the probability interval is -$1,300). However, this is by no means guaranteed: The upper end of the interval is $3,400.

5. EMISSION BENEFITS OF VEHICLES THAT SATISFY ZEV PROGRAM REQUIREMENTS

Section 4 examines the incremental costs of the various types of vehicles that may be produced to satisfy the ZEV program requirements. This section examines what these additional costs will buy in terms of emission reductions. As we did with costs, we calculate emission reductions relative to a vehicle that meets the super ultra low emission vehicle (SULEV) exhaust standard and the near-zero evaporative emission standard. This is the cleanest vehicle required under CARB regulations outside the ZEV program.

5.1 VEHICLE EMISSIONS

Motor vehicle emissions can be divided into three categories: tailpipe (or exhaust), evaporative, and indirect (or fuel-cycle). For each, we present estimates of the emission rates in grams per mile for the different advanced technology vehicle categories created by the ZEV program.

Tailpipe Emissions

Tailpipe, or exhaust, emissions include those occurring during vehicle start-up and warm-up and during operation at running temperature. Our estimates of tailpipe emission rates are taken from CARB's analysis of vehicle emissions using its most recent emission model—EMFAC2000 (CARB, 2000b, pp. 134-137). Table 5.1 lists average lifetime emissions for the advanced technology vehicles: the partial zero emission vehicle (PZEV); the advanced technology PZEV (ATPZEV), which in this case is a gasoline hybrid electric vehicle (GHEV); and the battery-powered electric vehicle (BPEV). The first set of rows in the table reports CARB's estimated average rates of tailpipe emissions in grams per mile over the life of the vehicle. Though not explicitly stated in CARB's write-up, these estimates are probably averages over the vehicle's "useful life," which CARB defines as 10 years and 120,000 miles (CARB, 2001g, p. C-1). As discussed in Subsection 4.4, average life and miles traveled are likely longer. CARB's emission rates assume different emission control system deterioration rates over the life of the vehicle. The second column of Table 5.1 lists emission rates with a high deterioration rate; the third column lists those with the standard deterioration rate used in past analyses. We use the higher rate for our estimate of SULEV emissions because emission rates based on the standard deterioration rates likely understate emissions over the full life of the vehicle.

Table 5.1

Average Lifetime Emissions of Advanced Technology Vehicles
(average grams per mile over life of vehicle)

Type of Emissions	SULEV, Near-Zero Evap., High Deterioration Rate	SULEV, Near-Zero Evap., Standard Deterioration Rate	PZEV	ATPZEV (GHEV)	BPEV
Tailpipe[a]					
NMOG	0.015	0.007	0.007	0.007	0
NOx	0.030	0.025	0.024	0.024	0
Evaporative[a]					
NMOG	0.032	0.032	0.020	0.020	0
Indirect[b]					
NMOG	0.031	0.031	0.031	0.021	0.002
NOx	0.016	0.016	0.016	0.011	0.003
Total					
NMOG	0.078	0.070	0.058	0.048	0.002
NOx	0.046	0.041	0.040	0.035	0.003
NMOG+NOx	0.124	0.111	0.098	0.083	0.005

[a]Estimates are from CARB, 2000b, pp. 134-137.
[b]Estimates are from Unnasch, Browning, and Montano, 1996.

Since PZEVs and ATPZEVs must meet SULEV tailpipe standards, CARB places their emission rates at the same level as a SULEV with the standard deterioration rate. We think this reasonable, because PZEVs and ATPZEVs must be warranted to 15 years and 150,000 miles. ZEVs, of course, have no tailpipe emissions.

Evaporative Emissions

Evaporative emission standards are set for diurnal, hot-soak, resting, and running-loss emissions.[1] The current standards for diurnal, hot-soak, and resting emissions have been in place since 1995 and require that emissions be less than 2 grams during a prescribed test procedure. CARB has adopted more stringent evaporative emission standards for all light-duty vehicles (LDVs). This near-zero evaporative standard requires emissions to be less than 0.5 grams during the prescribed test procedure and will be phased in between 2004 and 2006. The zero-evaporative emission standard for PZEVs and ATPZEVs, though not literally zero, requires evaporative emissions to be less than 0.25 grams during the test procedure. The standard for

---

[1]Diurnal emissions are associated with the diurnal breathing of the fuel tank as the ambient air temperature rises and falls. Hot-soak emissions occur when the fuel system is exposed to high under-hood temperatures after the vehicle is turned off. Resting losses occur when the vehicle is turned off and the hot-soak period has passed. Running losses occur when the vehicle is in operation.

running losses is 0.05 grams per mile and is not currently scheduled to change. PZEVs are not required to meet a stricter running-loss standard.

CARB's estimates of evaporative emissions for vehicles certified to the near-zero and zero standards are given in Table 5.1. CARB expects vehicles certified to the near-zero standard to emit 0.032 grams per mile of NMOG on average over their useful lives, and it expects vehicles certified to the zero-standard to emit 0.02 grams per mile. Zero-emission vehicles have no fuel-related evaporative emissions.[2]

Indirect Emissions

Indirect, or fuel-cycle, emissions are those produced during creation and distribution of the fuel that powers the vehicle. They comprise all emissions up to the point at which the fuel is *in* the vehicle, which means those emissions produced during

- extraction (feedstock extraction and transport)
- production (e.g., oil refining or electricity generation)
- marketing (e.g., fuel bulk storage, transportation, transmission, gaseous fuel compression)
- distribution (i.e., emissions from local fuel stations and from the process of vehicle refueling).

We use CARB's estimates of indirect emissions (CARB, 2000b, pp. 135-137), which are drawn from Unnasch, Browning, and Montano, 1996, and which predict indirect emissions in the South Coast Air Basin in 2010. These estimates reflect emission controls and vehicle fuel economy expected in 2010. The estimates for ATPZEVs are lower than those for SULEVs and PZEVs because GHEVs have better gas mileage.[3] Indirect emissions for BPEVs are small but non-zero due to emissions from electricity generation in the South Coast.[4]

CARB has not analyzed the indirect emissions associated with hydrogen generation. As discussed in Subsection 4.7, we have assumed that hydrogen is generated at filling stations from natural gas piped to the sites. There are some carbon monoxide emissions associated with the reforming process itself (Ogden, 2001, p. 9). Energy is required to run the reformer. If methane

---

[2]ZEVs do have some nonfuel NMOG emissions (from plastic parts inside the vehicle, for example), but we assume that these are the same as in non-ZEVs.

[3]Indirect emissions are related to fuel economy. While CARB is not explicit about why its indirect emission estimates for ATPZEVs are lower, the explanation may be a 50 percent increase in fuel economy.

[4]CARB's estimates for indirect omissions are based on the fuel economy of vehicles sold today. Congress recently declined to raise fuel efficiency standards but directed the Executive branch to consider new standards. Higher fuel economy will mean that the incremental emission reductions of BPEVS will be lower. It may also mean lower incremental emission reductions for GHEVs (unless their fuel economy rises proportionately).

is burned to provide this energy, some NOx emissions will result. If electricity is used, then there will be NOx and NMOG emissions at the power plant. Electricity will also presumably be used to compress the hydrogen to 5000 psi, inducing additional indirect emissions. We have not been able to develop estimates of the indirect emissions associated with direct hydrogen fuel-cell vehicles (DHFCVs) relative to those for BPEVs. For lack of anything better, we set indirect emissions of DHFCVs equal to those of BPEVs. Further investigation of the indirect emissions of DHFCVs is warranted.

5.2 VEHICLE EMISSIONS IN PERSPECTIVE

In percentage terms, ZEVs are dramatically cleaner than SULEVs and even PZEVs and ATPZEVs. Figures 5.1 and 5.2 show the total lifetime emissions in pounds when estimates of the average rates of emissions in grams per mile are multiplied by the assumed life of the vehicle.[5,6] A BPEV's NMOG emissions are 97 percent lower than those for a vehicle meeting the SULEV exhaust (using the high deterioration rate) and near-zero evaporative emission standards; its NOx emissions are 94 percent lower. These comparisons obscure the fact that SULEVs with near-zero evaporative emissions are quite clean, however, and that absolute emission reductions are not very large. A SULEV with near-zero evaporative emissions will emit only 25.8 pounds of NMOG and 15.2 pounds of NOx over a 150,000-mile lifetime.

To put the emission reductions of ZEVs in perspective, we project LDV emissions in the South Coast Air Basin in 2010 assuming that the entire LDV fleet is composed of each vehicle type in Table 5.1. It is not realistic to think that all vehicles will be certified to one of these standards in 2010, but such a comparison gives an idea of the effect of moving from a fleet of the cleanest types of vehicles required under current regulations (SULEVs with near-zero evaporative emissions) to a zero emission fleet. To estimate total emissions with each of the different emission standards, we multiply CARB's forecast of miles traveled by LDVs—i. e., passenger cars and light-duty trucks (LDTs)—in the South Coast in 2010 by the estimates of vehicle emissions per mile.[7]

Table 5.2 shows that resulting emission reductions are not large relative to the carrying capacity of the South Coast Air Basin. The table shows that if the entire LDV fleet in 2010 were SULEVs with near-zero evaporative emissions, its emissions would be 30 tons of NMOG and

---

[5]For these figures, we assume each vehicle travels 150,000 miles on average over its lifetime.

[6]Emissions are not discounted back through time here. They are, however, discounted when we calculate cost-effectiveness.

[7]CARB's EMFAC2000 predicts that LDVs will travel 343 million miles a day in the South Coast in 2010.

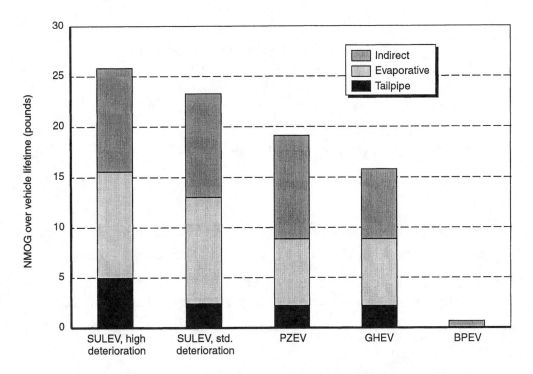

Figure 5.1—Average Vehicle Lifetime NMOG Emissions

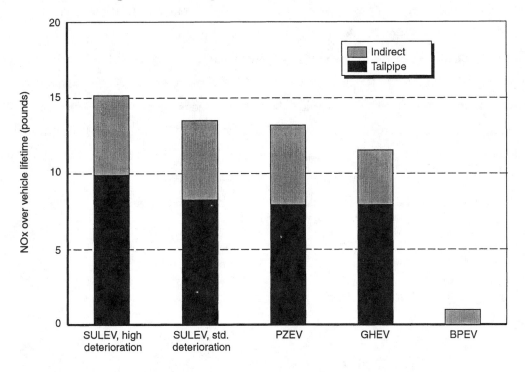

Figure 5.2—Average Vehicle Lifetime NOx Emissions

Table 5.2

**Light-Duty Vehicle Fleet Emissions When Entire Fleet Is
Certified to Various Emission Standards
(tons per day)**

	NMOG	NOx	NMOG+NOx
SULEV, near-zero evap. emissions, high deterioration	30	17	47
SULEV, near-zero evap. emissions, standard deterioration	26	16	42
PZEV	22	15	37
ATPZEV (GHEV)	18	13	31
BPEV	1	1	2

17 tons of Nox per day. Moving the entire fleet to BPEVs would reduce daily NMOG emissions by 29 tons and daily NOx emissions by 16 tons. Recall from Subsection 2.2 that the carrying capacity of the South Coast Air Basin is 401 tons per day for NMOG and 525 for NOx. Moving from SULEVs with near-zero evaporative emissions to ZEVs would thus eliminate emissions equivalent to only 7 percent of the basin's carrying capacity for NMOG and 3 percent of it for NOx.

As we discussed in Subsection 2.3, California is far short of the reductions in NMOG emissions it needs to meet air quality standards by 2010. Programs must still be found that reduce NMOG emissions in the South Coast by 138 tons per day. The target for NMOG emissions from LDVs in 2010 in the South Coast Air Quality Management District plan is 55 tons per day (see Table 2.1). Requiring the entire LDV fleet to meet the SULEV and near-zero evaporative emission standards would reduce NMOG emissions from 55 to 30 tons per day. The 25-ton reduction could be credited against the 138-ton shortfall. Going to a ZEV fleet would take an additional 29 tons out of the shortfall. But California must ask whether eliminating the last 29 tons of LDV emissions is a cost-effective approach.

We now turn to the cost-effectiveness of advanced technology vehicles.

6. COST-EFFECTIVENESS OF ZEVS AND PARTIAL ZERO EMISSION VEHICLES

We begin this section by combining the cost estimates in Section 4 with the emission reductions in Section 5 to predict the cost-effectiveness of technologies that manufacturers may use to meet ZEV program requirements. Cost-effectiveness is measured as the cost per ton of non-methane organic gases (NMOG) plus oxides of nitrogen (NOx) emissions reduced. We then compare these estimates with the cost-effectiveness of other measures that have recently been adopted or are scheduled to be adopted to reduce emissions of ozone precursors in the South Coast. Finally, since cost-effectiveness calculations rarely cover all considerations that should be taken into account when developing policies and programs, we examine potential costs and benefits of California's ZEV program that are not reflected in our cost-effectiveness estimates.

6.1 COST-EFFECTIVENESS ESTIMATES

We first calculate the cost-effectiveness of the various advanced vehicle technologies in the near term. For battery-powered electric vehicles (BPEVs), advanced technology partial zero emission electric vehicles (ATPZEVs), and partial zero emission vehicles (PZEVs), this period is 2003 through 2007. For direct hydrogen fuel-cell vehicles (DHFCVs), it is 2006 through 2010, because we think it unlikely that fuel-cell vehicles will be ready for the market in more than pilot quantities before then (see Subsection 3.1). We then project cost-effectiveness when the vehicles are in volume production.

The low costs in volume production can be thought of as the payoff society can expect after incurring high costs when a new technology is introduced. But in evaluating whether an investment makes sense, society must consider not only the payoff, but also the costs of achieving it. Thus, we also calculate cost-effectiveness for each technology that includes the higher costs of vehicles produced prior to volume production.

We calculate the cost-effectiveness of moving to progressively tighter emission control standards. We first calculate the cost-effectiveness of a PZEV relative to a vehicle that meets the super ultra low emission vehicle (SULEV) exhaust and the near-zero evaporative standards. Then we calculate the cost-effectiveness of ATPZEVs, which in this case are gasoline hybrid electric vehicles (GHEVs), and ZEVs relative to PZEVs. This allows us to evaluate whether the PZEV standard is sensible on cost-effectiveness grounds and, in turn, whether it makes sense to further reduce vehicle emissions.

Our estimates of cost-effectiveness implicitly assume that consumers are willing to pay the same amount, on a lifecycle basis, for the vehicles being compared. It assumes, for example, that consumers value a full-function electric vehicle (EV) with a 100-mile range just as much as they value a comparably sized internal combustion engine vehicle (ICEV) with the same lifecycle cost. Consumers may be willing to pay more for a full-function EV if they view the benefits of a quiet ride, home recharging, and low emissions as outweighing the drawbacks of limited range. If they are willing to pay more, then this part of the higher full-function EV cost should not be included in the cost-effectiveness calculation. But consumer willingness to pay more for EVs is a subject of great debate. Appendix E reviews studies of the market for EVs. Based on our review, we think it reasonable to compare vehicles with the characteristics specified here based on their lifecycle costs.

Cost-Effectiveness in Near-Term and in Volume Production

To calculate cost-effectiveness for vehicles produced between 2003 and 2007, we divide the 90 percent probability intervals for discounted incremental lifecycle cost by the emission reductions generated over the lifetime of the vehicle. The result is cost per ton of emissions reduced, or the *cost-effectiveness ratio*. The higher the cost per ton, the less cost-effective the technology. Emissions are discounted back to the time of the vehicle sale. Discounted incremental lifecycle costs are taken from the analysis described in Subsection 4.4; they include the incremental costs of producing the vehicle and the incremental operating and maintenance costs over the life of the vehicle discounted back to the time of production. We assume that vehicle mileage (and thus emission reductions) is spread evenly over the vehicle's lifetime. All vehicles except city EVs are assumed to travel 10,000 miles per year for 15 years. City EVs are assumed to travel 5,000 miles per year for 15 years. As with costs, emission reductions are discounted back to the time of vehicle production using a 4 percent discount rate.[1]

Cost per ton in volume production is calculated in the same manner. The discounted emission reductions are identical to those for vehicles sold between 2003 and 2007. The discounted incremental lifecycle costs are taken from Table 4.17 (see Subsection 4.7).

Table 6.1 reports the results of the analysis. The brackets in each cell contain the interval into which cost per ton is likely to fall given our estimates of emission reduction and the 90 percent probability intervals for incremental lifecycle cost.[2] For example, we predict that cost per ton for full-function EVs with NiMH batteries produced between 2003 and 2007 will be from $1.5 to $2.3 million per ton of NMOG plus NOx reduced. In volume production, the cost per ton

---

[1]The discounting of emission benefits is controversial among noneconomists, but is standard in economic analysis.

[2]Because our emission reduction estimates are point estimates (constants), the intervals for cost per ton are still formally 90 percent probability intervals. Of course, there is uncertainty in the emission reduction estimates, but we have not attempted to characterize it in this analysis.

Table 6.1

Cost-Effectiveness Ratio of Vehicle Technologies in Near Term and in Volume Production (intervals in brackets; $1000s per ton of NMOG+NOx reduced)

Vehicle Technology	Near Term	High-Volume Production
	2003-2007	
PZEV relative to SULEV with		
near-zero evap. emissions	[18, 71]	[18, 71]
GHEV relative to PZEV	[650, 1,800]	[-500, 86]
Full-function EV relative to PZEV		
NiMH	[1,500, 2,300]	[260, 710]
PbA	[750, 1,500]	[12, 400]
City EV relative to PZEV		
NiMH	[1,400, 2,100]	[70, 370]
PbA	[850, 1,600]	[-80, 160]
	2006-2010	
DHFCV relative to PZEV	[650, 1,300]	[-110, 270]

of emissions reduced will likely still range from $260,000 to $710,000. These full-function EV costs and emission reductions are relative to those of a PZEV. We predict that relative to a SULEV with near-zero evaporative emission control, a PZEV will reduce emissions at a cost of $18,000 to $71,000 per ton.[3]

Cost-Effectiveness Including Transition Costs

To calculate cost-effectiveness for each technology that includes the costs and emission benefits of vehicles produced before costs fall to volume production levels, we specify production volumes and the decline in lifecycle cost over time. For the production volumes, we take the midpoint of the production scenarios we developed for each vehicle type in Subsection 3.2. Incremental lifecycle costs in 2003 are alternately set at the upper and lower bounds of the ranges for 2003 to 2007 reported in Table 4.17 and then fall to their high-volume levels by the dates shown in Table 6.2. For all vehicles except PZEVs, we selected these dates because our estimated cost-volume relationships (see Section 4) imply that annual production volumes will be high enough by these dates for costs to be approaching high-volume levels. For PZEVs, we assume that incremental costs remain unchanged from 2003 on. The decline in costs between 2003 and the year costs reach their high-volume levels for vehicles other than PZEVs reflects the decline in costs predicted by the cost-volume relationship.[4] We discount both costs and emission

---

[3]In some cases, the cost per ton is negative. This occurs when the discounted incremental cost is negative.

[4]In particular, we assume that the cumulative percent decline predicted by the cost-volume relationship between 2003 and the year costs reach their high-volume levels matches the cumulative percent decline in the costs used here between 2003 and the year costs reach their high-volume levels.

Table 6.2

**Year in Which Incremental Lifecycle Costs
Fall to High-Volume Predictions**

Vehicle Technology	Year High-Volume Cost Achieved
GHEV	2015
Full-function EV	
NiMH	2015
PbA	2015
City EV	
NiMH	2015
PbA	2015
DHFCV	2020

reductions back to 2002, and we run the simulation out to the year 2100. We chose 2100 because at a 4 percent discount rate, discounted costs and emission reductions beyond 2100 are not large (in 2100, costs and emission reductions are multiplied by 0.021 to bring them back to 2002). Production volumes are fixed at their 2030 levels (the last year projected in Subsection 3.2) through 2100.

Table 6.3 reports the effect of including transition costs in the calculation of cost-effectiveness. The brackets contain the lower and upper bounds of our estimates.[5] The second and third columns show the total discounted costs and discounted emission reductions for the assumed volume scenario. The last column reports cost per ton of NMOG plus NOx reduced in current dollars. As expected, the cost-effectiveness ratio is higher when transition costs are included than when only high-volume costs are considered. For example, cost per ton ranges from $12,000 to $400,000 per ton in Table 6.1 for a full-function EV with PbA batteries in volume production. When we include the costs incurred before costs fall to their high-volume levels, however, cost per ton rises to $50,000 to $470,000 per ton. The resource costs generated under these different scenarios are large. Discounted incremental costs of full-function EVs with NiMH batteries add up to $5.3 to $13.1 billion, yet the vehicles generate only 16,000 tons of discounted emission reductions.

6.2 COST-EFFECTIVENESS OF ADVANCED VEHICLE TECHNOLOGIES IN PERSPECTIVE

Subsection 2.4 reviews the cost-effectiveness of measures recently adopted or being considered to reduce NMOG and NOx emissions in the South Coast Air Basin. We found that regulations to reduce NMOG emissions from consumer products have cost up to $7,000 per ton

---

[5]Even though based on the 90 percent probability intervals, these ranges are, strictly speaking, no longer 90 percent probability intervals.

Table 6.3

Cost-Effectiveness Ratio of Vehicle Technologies Including Transition Costs (ranges in brackets)

Vehicle Technology	Discounted Costs ($millions)	Discounted Emission Reductions (1000s of tons NMOG+NOx)	Cost per Ton ($1000 per ton NMOG+NOx reduced)
PZEV relative to SULEV with near-zero evap. emissions	[850, 3,300]	47	[18, 71]
GHEV relative to PZEV	[-1,500, 600]	3	[-440, 180]
Full-function EV relative to PZEV			
NiMH	[5,300, 13,100]	16	[330, 810]
PbA	[870, 7,700]	16	[50, 470]
City EV relative to PZEV			
NiMH	[1,200, 3,800]	8	[150, 470]
PbA	[-200, 2,000]	8	[-20, 250]
DHFCV relative to PZEV[a]	[-1,300, 5,000]	16	[-80, 310]

[a]Year 2006 and on, only.

reduced and that regulations to reduce NMOG emissions from other stationary sources are expected to cost up to $25,000 per ton. Regulations on diesel engines target mainly NOx emissions and have been inexpensive on a cost-per-ton basis—all $800 per ton or less. Further regulation of gasoline engines is more expensive. Estimates of reducing light-duty vehicle (LDV) emissions by scrapping old vehicles or through the enhanced Smog Check program run up to $33,000 per ton of NMOG plus NOx reduced, and regulation of large and small off-road gasoline engines reaches nearly $21,000 per ton.

While these estimates are sometimes higher than the guidelines that the South Coast Air Quality Management District (SCAQMD) and CARB have set to evaluate the cost-effectiveness of proposed regulations,[6] they are at least comparable to dollar estimates of the benefits of reducing NMOG and NOx emissions. In their review of the literature, Dixon and Garber (1996, pp. 21-22, 364-370) conclude that the benefits of reducing NMOG and NOx emissions in the South Coast likely exceed $5,000 per ton, perhaps by a substantial amount, but are probably less than $25,000 per ton.[7] (These figures, which are in 1995 dollars, rise to $5,800 and $29,000 when converted to 2001 dollars.[8])

---

[6]Cost-effectiveness greater than $13,500 per ton of NMOG triggers further review at SCAQMD. CARB guidance sets an upper limit of $22,000 per ton of NMOG plus NOx (see Subsection 2.4).

[7]The studies vary by the extent to which they include chronic versus acute human health effects, damage to plants and animals, and damage to materials. The range in benefit estimates is in part due to this variation.

[8]Conversion done using the consumer price index. See http://data.bls.gov/cgi-bin/surveymost.

On a cost-per-ton basis, PZEVs appear to be an attractive way to reduce NMOG and NOx emissions. The low end of our predicted range ($18,000) is comparable to the costs of many recently adopted regulations per ton. The upper end of the range ($71,000) is substantially higher than the cost of recently adopted regulations. However, it seems at least plausible that cost per ton will have to rise to such a level if emission reduction targets are to be met in the South Coast. The cost-effectiveness estimates for PZEVs are based on the costs of converting passenger cars and the smallest light-duty trucks (CARB's LDT1 category) to PZEVs. Costs of converting the heavier pickups, sport utility vehicles (SUVs), and minivans common on the road today (CARB's LDT2 category) to PZEVs may be higher.

Moving from PZEVs to GHEVs may or may not be a sensible way to reduce ozone levels. When GHEV maintenance costs are comparable to those of ICEVs, the cost per ton of emissions reduced is low or even negative. However, if the battery must be replaced during the life of the vehicle, the cost per ton is not particularly attractive, especially when transition costs are included. Because the difference between PZEV and GHEV emissions is small, cost per ton is very sensitive to changes in the incremental cost of GHEVs. GHEV incremental cost must therefore be small if cost per ton is to be moderate.

The cost per ton of moving from PZEVs to full-function EVs with NiMH batteries is huge. Even in volume production, the estimates run from $260,000 to $710,000 per ton of NMOG plus NOx reduced. Full-function EVs with PbA batteries look more attractive, but once transition costs are included, the cost is still $50,000 to $470,000 per ton (see Table 6.3). What is more, there are real doubts about whether it is practicable to produce full-function EVs with a 100-mile range and broad appeal using PbA batteries.

The high cost-effectiveness ratio reflects the small emission difference between PZEVs and BPEVs. The lower end of our range for the incremental high-volume cost of a full-function EV with PbA batteries is only $200 per ton (see Table 4.17), yet the lower end of the cost-effectiveness range becomes $50,000 per ton once transition costs are included (see Table 6.3). This means that the lifecycle costs of ZEVs must be quite close to those of PZEVs if costs per ton are not to rise to high levels (say, more than $50,000 per ton).

Emission reductions from city EVs using NiMH batteries cost less than those from full-function EVs with NiMH batteries do, but the cost per ton is still high. Including transition costs, estimated cost per ton runs from $150,000 to $470,000. In Subsection 4.7, we conclude that the lifecycle costs of city EVs with PbA batteries may be less than those of small ICEVs in volume production. In this case, the cost per ton would be negative, even including transition costs (-$20,000 per ton). Two important caveats are warranted, however. First, most manufacturers are not using PbA batteries in their city EVs, which suggests that there are important design or performance disadvantages to using PbA batteries in city EVs. Second, there is some chance that the cost per ton for a city EV with PbA batteries will be low, but there is also a good chance that it will be quite high. The low end of the range is based on optimistic assumptions about the parameters that determine lifecycle costs, but the upper end of the range ($250,000) is based on parameter values that are no less likely.

DHFCVs will be very expensive on a cost-per-ton basis when they are first introduced, but they may be a sensible strategy for reducing the emissions of ozone precursors in the long run. We project that in high-volume production, the lifecycle cost of a DHFCV could be less than that of an ICEV. However, a great deal of uncertainty about high-volume DHFCV cost remains: If it turns out to be at the top of the estimated range, it would rise to $310,000 per ton of emissions reduced once transition costs are included. What is more, widespread use of DHFCVs—a precondition for the volume production needed to achieve high-volume costs—depends on the availability of a hydrogen fueling infrastructure. Little progress has been made in developing such an infrastructure, and its development is hardly assured.

6.3 ADDITIONAL FACTORS TO CONSIDER WHEN EVALUATING ZEV PROGRAM

Our cost-effectiveness estimates do not capture all potential costs and benefits of the vehicles that manufacturers may use to satisfy ZEV program requirements. Here we discuss several additional considerations that might enter into decisions on whether to modify the program. We first examine several of the program's uncounted potential benefits:

- technology forcing
- job creation
- insurance against disappointments in ICEV emission performance
- reduced dependence on foreign oil
- reductions in vehicle emissions other than NMOG and NOx.

We then consider several uncounted potential costs:

- decreases in new vehicle sales
- increases in vehicle miles traveled
- emission reductions in all areas rather than only where there are air quality problems
- offsetting interactions with other programs.

These factors will tend to increase the estimates of cost per ton that we have presented so far.

Uncounted Potential Benefits

Technology Forcing. Our estimates of costs in volume production are based on what is currently known about EV technology. It may be that requiring automakers to manufacture large numbers of EVs will lead to unanticipated technological advances that will substantially reduce the cost per ton of emissions reduced. Such "technology-forcing" regulation is often held to be an effective means of overcoming market failures (e.g., the failure to consider pollution costs in designing vehicles) because it compels innovation that would not otherwise occur. In evaluating the ZEV program, policymakers need to assess the likelihood that the program will produce unforeseen breakthroughs in EV technology.

Indeed, since their inception in the 1960s,[9] most automobile emission regulations have been technology forcing (Brown et al., 1995). The first major federal regulation, Title II of the Clean Air Act of 1970, required 90 percent reductions in hydrocarbon and carbon monoxide emissions by 1975 and in NOx emissions by 1976—standards that were "a function of the degree of control required, not the degree of technology available at the time."[10] Domestic automakers fought the regulations, calling them technically infeasible and ruinously expensive.[11] The technology-forcing provisions of subsequent regulations, such as the Corporate Average Fuel Economy (CAFE) standards, the California LEV I program, and the 1990 Clean Air Act Amendments, have met with similar resistance from domestic automakers.

Despite the resistance, however, the automotive industry has made remarkable progress in emission control. Against this backdrop, a CARB official responding to an automaker's objection to the ZEV program recently declared that "projections of future technologies must be viewed in the context of these remarkable advancements that have been achieved by the auto industry" (CARB, 2001f). The thinking is, if technology-forcing regulations on ICEVs have induced important technological breakthroughs, why should the ZEV program not do the same?

Even though regulation has induced development of ICEV emission control technologies, it does not necessarily follow that automakers will uncover technological breakthroughs that will make BPEV's commercially viable. First of all, despite more than $500 million having been spent on EV battery development since 1995 by battery makers, the automobile industry, and the

---

[9]California enabled the promulgation of automobile emission controls in the Motor Vehicle Pollution Control Act of 1960, with the first regulations taking effect in 1965. Federal regulations were enabled by the Motor Vehicle Air Pollution Control Act of 1965 and were implemented in 1968 (Percival et al., 1992).

[10]S. Report No. 91-1198, 91st Cong., 2nd Sess. 24 (1970). A Senate staffer noted that the 90 percent figure was "a back of the envelope calculation. . . . We didn't have any particular methodology. We just picked what sounded like a good goal" (Easterbrook, 1989).

[11]A GM vice president warned that outfitting its fleet with catalytic converters raised the "prospect of an unreasonable risk of business catastrophe . . . [and] complete stoppage of the entire production." A Ford official claimed that it would "cause Ford to shut down and would result in: 1) reduction of gross national product by $17 billion; 2) increased unemployment of 800,000; and 3) decreased tax receipts of $5 billion at all levels of government so that some local governments would become insolvent" (California ZEV Alliance, 2001).

United States Advanced Battery Consortium (Anderman, Kalhammer, and MacArthur, 2000), full-function EVs will still have limited range and will cost substantially more than ICEVs for the foreseeable future. The Battery Technology Advisory Panel (BTAP) found that major technological advances would be required to reduce costs significantly and that they were unlikely for the next six to eight years (Anderman, Kalhammer, and MacArthur, 2000). In spite of much ingenuity and resolve, the sought-after dramatic breakthrough just has not happened.

Second, the context of technology forcing in the ZEV program is different from that of the ICEV regulations. The ICEV regulations set performance standards for fleet emissions and leave it to automakers to figure out how to meet them. The ZEV program in effect tells manufacturers that they have to meet part of their fleet NMOG requirement with a special type of vehicle—a ZEV. The ZEV program is thus mandating a particular approach to reducing vehicle emissions.

Even if BPEVs turn out to be a dead end, it can be argued that the ZEV program accelerated the development of the motors, controllers, and advanced batteries that may make fuel-cell vehicles economical. Research on BPEVs likely also sped the development of GHEVs.[12] However, there is a risk that the ZEV program is forcing technology in the wrong direction. It may be that the most cost-effective way to reduce emissions is to build vehicles that have very low, but not zero, emissions. Forcing companies to build ZEVs may divert attention from more-promising areas, such as more-effective control technologies for heavier light-duty trucks (LDT2s) or for consumer products and other stationary sources.

Thus, it does not make sense to proceed with the ZEV program solely based on the conviction that the technologies that make ZEVs attractive will materialize. The risk is that a large social investment in building thousands of high-cost vehicles will yield little or no return and that other, more-attractive options will be overlooked.

Job Creation. The ZEV program has compelled automakers and their suppliers to engage in substantial research and development (R&D) to meet future standards, and these R&D expenditures yield a variety of economic benefits through the creation of R&D jobs and the attendant multiplier effects. Any such benefits are not captured in the cost-effectiveness estimates presented above.

Many California companies, including the state's aerospace giants, have experience in the advanced materials and microelectronics that EVs require, and the Los Angeles area is home to many specialty and aftermarket auto component companies.[13] A regional policy center found

---

[12]For an examination of the so-called secondary benefits of ZEVs, see Burke, Kurani, and Kenney, 2000.

[13]EVs require a large number of components and manufacturing processes not shared with conventional vehicles and, at least in the initial years of the program, will be built in small batches rather than in continuous assembly lines.

over 400 companies in the Los Angeles area with the capability to manufacture EVs (Reinhold, 1992). A CALSTART survey identified over 100 companies in California with ties to EV R&D and manufacturing, and the great majority of survey respondents reported that the ZEV program was instrumental to their firm's existence and continued growth (Burke, Kurani, and Kenney, 2000). CALSTART estimated that in 2000 there were approximately 3,500 employees at these EV-related companies, with about 800 of their jobs attributed to the program. An analysis of the potential for job creation in Los Angeles estimates that if 10 percent of the cars sold in California were EVs produced in Los Angeles County, over 24,000 jobs would be added to the local economy. If the region specialized in EV components rather than assembly, approximately 10,000 jobs would be added (Wolff et al., 1995). A 1994 CARB staff report predicted that the ZEV program would yield significant economic growth in California (55,000 new jobs by 2000, and 70,000 by 2010) because 70 percent of the parts in a ZEV would be different from those in ICEVs (Reed, 1997). The California Electric Transportation Coalition predicted that battery manufacturing alone could generate 10,000 jobs in the state by 2005 (Brown et al., 1995).

There are many reasons to be skeptical about the supposed economic benefits of the ZEV program. It was formulated in the early 1990s, when California manufacturing was mired in a recession and the aerospace industry was hard hit by the decline in defense spending after the Cold War ended. The above estimates of job generation are less credible (or less meaningful) in periods of full employment, as have been enjoyed in the state in recent years. Even if there are net employment gains in California, jobs might be created at the expense of workers in traditional automotive industries elsewhere in the country.

Furthermore, a shift to EVs may actually yield a net decrease in *total* auto manufacturing employment, since EVs have far fewer parts and are simpler to assemble than conventional automobiles. Wolff et al. (1995) found that every $1 billion in final demand for EVs removes about 4,000 jobs, and that if EVs were to replace all conventional vehicles, nearly 300,000 jobs would be lost. Studies finding very high manufacturer costs for compliance with the ZEV program (the so-called "ZEV tax") also find an associated reduction in employment levels in California. One such report, by National Economic Research Associates (2000), estimates that the program, as revised in January 2001, will result in a loss of 10,000 jobs by 2020.

Insurance Against Disappointments in ICEV Emission Performance. The ZEV program also provides some insurance against the possibility that the lifetime emissions of PZEVs and other very clean ICEVs do not turn out to be as low as currently projected. This is another benefit that is difficult to quantify. PZEVs have yet not been sold, and it will be many years before their in-use emissions can be verified. It may turn out that the on-board diagnostics (OBD) systems that monitor emission performance do not work well when emissions are at such low levels, or that emission control systems deteriorate faster than expected. The ZEV program does provide some insurance, but the question is, at what cost. Other types of insurance may be more cost-effective—e.g., research on how to further reduce emissions from stationary sources and off-road and diesel vehicles.

Reduction of Dependence on Foreign Oil. The U.S. dependence on foreign oil, and particularly on oil from the Middle East, concerns many policymakers because of the potential uncertainty in supply and because of the cost of maintaining security forces abroad to decrease that potential uncertainty. Greater use of BPEVs, DHFCVs, and GHEVs entails less gasoline use and thus, presumably, lower imports of foreign oil. BPEVs lower gasoline use because most of the electricity that would charge EVs in California would be generated using coal, natural gas, nuclear energy, or hydro power. DHFCVs lower gasoline use because the hydrogen that powers fuel cells would likely be generated from natural gas, not oil. GHEVS lower gasoline use because they are more energy efficient.

To provide a rough estimate of how the ZEV program might affect oil use in California, we assume that manufacturers satisfy the program by producing as many PZEVs and GHEVs as allowed and satisfy the ZEV proportion of the program with full-function EVs (see Subsection 3.2 for a discussion of the various production scenarios). We then calculate the amount of gasoline displaced annually and the barrels of oil displaced, assuming one barrel of oil yields 19.5 gallons of gasoline.[14] As can be seen in Figure 6.1, the amount of oil displaced rises from roughly 250,000 barrels a year in 2003 to nearly 2 million barrels a year in 2030.

To put these reductions in perspective, consider that almost 7 billion barrels of oil are consumed annually in the United States, with California accounting for approximately 10 percent of the total (California Academy of Sciences, 2001). The United States imports approximately 950 million barrels of oil a year from the Middle East (Energy Information Administration, 2001). Thus, even at 2 million barrels a year, the amount of oil displaced by the ZEV program is small relative to total oil usage in California and total oil imports from the Middle East. Of course, the amount of oil displaced will rise if the prevalence of ZEVs or hybrids increases in the state and as they spread to other parts of the country. For example, if EVs had accounted for one-quarter of the miles driven statewide in 2000, oil use would have been nearly 100 million barrels lower.[15] A high enough penetration of ZEVs and hybrids in California and other states would eliminate the need to import oil from the Middle East.

---

[14]Figures taken from the American Petroleum Institute, 2001.
[15]The number of vehicle miles traveled in 2000 was 300 billion.

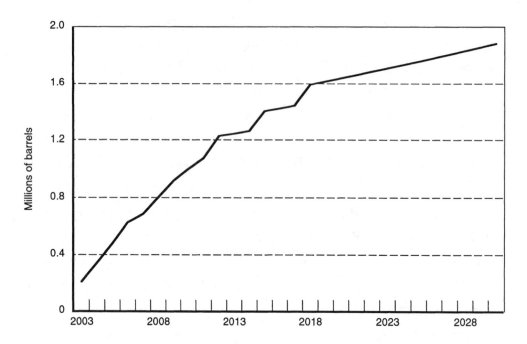

Figure 6.1—Barrels of Oil Displaced Annually by ZEV Program

Although difficult to quantify, there is a monetary cost connected with maintaining the flow of oil from the Middle East. According to the U.S. Department of Energy,

> [The cost of m]aintaining the uninterrupted flow of oil from the Gulf region is high—as much as $57 billion per year. The U.S. General Accounting Office estimated that the cost of U.S. military and foreign aid programs in the Gulf area from 1980 to 1990 was as high as $365 billion. When military and energy security factors are taken into consideration, the true cost of oil is as high as $100 per barrel or $5 a gallon. . . . And this doesn't take into consideration the cost of actual military action to defend our interests in the Persian Gulf. Our most recent experience with this was the Persian Gulf War which cost $61 billion and loss of priceless human lives (DOE, 2001).

These figures suggest that the gains from reducing Middle East oil imports are high. But these gains probably cannot be realized unless oil imports from the Middle East are substantially or completely reduced, since slight reductions would most likely have little effect on military and security expenditures in this area.

The ZEV program may thus reduce gasoline use and perhaps dependence on Middle East oil. However, the benefits most likely would be great only if the United States ceases importing substantial amounts of Middle East oil. Also, there may be much more cost-effective ways to reduce gasoline use—e.g., by increasing the Corporate Average Fuel Economy (CAFE) standard or by switching to alternative fuels, such as compressed natural gas, that produce low, but non-zero, emissions. Alternatives should be investigated before a great deal of weight is put on the ZEV program's potential to reduce dependence on foreign oil.

Reductions in Vehicle Emissions Other Than NMOG and NOx. Our analysis of cost-effectiveness considers reductions in the emissions of NMOG and NOx only. ZEVs will also reduce emissions of other pollutants such as particulates, carbon monoxide, and the greenhouse gas, carbon dioxide. In evaluating the importance of these other emission benefits, policymakers need to consider both the magnitude of the reductions and their value. For example, most areas of the state are in compliance with carbon monoxide standards, so additional carbon monoxide reductions may not be of great value. ZEVs may reduce carbon dioxide emissions, but there may be more cost-effective ways to achieve the same end. For example, it may be more cost-effective to raise the CAFE standard than to require ZEVs. The GHEVs recently offered for sale by Honda and Toyota suggest that raising the CAFE standard may not be as expensive or as demanding of lifestyle changes as previously thought.

Uncounted Potential Costs

Feedback on New Vehicle Sales. The ZEV program may cause manufacturers to increase the prices of conventional new vehicles in order to recover the costs of developing and producing ZEVs. Any such increase in new vehicle prices will tend to reduce the sales of new vehicles, slow the retirement of older vehicles, and gradually increase the age of the fleet (Gruenspecht, 1982a, 1982b). Older vehicles, because of deterioration and the less stringent emission requirements in force when they were manufactured, tend to have higher emissions than newer vehicles do. Thus, increases in new ICEV prices will tend to increase fleet emissions over time and to offset, at least to some extent, the emission benefits of ZEVs.

Whether the price increases will be large enough to overwhelm the emission benefits of ZEVs has been hotly debated. Harrison et al. (2001) argue that manufacturers will only increase the price of new vehicles sold in California and that those increases will be large enough to increase fleet emissions in California at least through 2020. CARB (2001e) counters that manufacturers will spread costs nationwide, and that even if they did not, the magnitude of the increase in California and the consequent effect on emissions would be much less than predicted by Harrison et al.

There are good arguments on both sides of this debate. The ZEV program does create a cost of selling an additional ICEV in states that have adopted the program.[16] Simple models of profit maximization conclude that manufacturers set prices on products according to the costs of producing and selling those products. The ZEV program creates no additional costs in states that

---

[16]For every ICEV sold, manufacturers must sell a fraction of a ZEV.

have not adopted the program, so prices should not rise in those states. Complications in the real world raise doubts about this reasoning, however. First, competition from small- and intermediate-volume manufacturers not subject to the pure ZEV portion of the program may dissuade the large-volume manufacturers from concentrating price increases in California. Now that the cutoff between intermediate- and large-volume manufacturers has risen to 60,000 vehicles per year (from 35,000 previously), large price increases by large-volume manufacturers may have real consequences for their market share. Second, manufacturers have spread costs outside the markets that generate them in a number of circumstances. Dixon and Garber (1996) were told by observers inside and outside the auto industry that companies typically spread vehicle transportation and delivery costs across geographic areas. The Green Car Institute found that manufacturers had recently dropped the $100 typically added to a vehicle's retail price to cover California emission requirements because "from a market standpoint the automakers viewed the separate charge for the California emissions programs as negative to their other marketing efforts" (Green Car Institute, 2001, p. 24). Manufacturers may be less likely to spread costs if the additional costs are large (as opposed to modest, as in the case of transportation and shipping charges); but in any case, uncertainty remains about the ZEV program's effect on new vehicle prices and any consequent indirect effect on fleet emissions in California.

Even if manufacturers spread costs nation- or even worldwide, there may be some reductions in new vehicle sales and, consequently, increases in emissions both inside and outside California.[17] Thus, consideration of the ZEV program's feedback on new vehicle sales would lead to an increase in the cost-per-ton estimates presented here, but the overall significance of the effect is uncertain.

Increase in Vehicle Miles Traveled. Studies have found that increases in fuel economy (or reductions in fuel price) increase the amount of driving done by households. For example, Greene, Kahn, and Gibson (1999) found that about 20 percent of projected energy savings due to increased fuel efficiency were offset by increases in miles driven. Such might be the case for the vehicles examined in our analysis. As shown in Table 4.17, the operation and maintenance costs of full-function EVs, city EVs, GHEVs, and DHFCVs may be less than those of a comparable standard ICEV in volume production. These lower costs may induce more driving and thus offset some of the emission benefits assumed. The result will be higher costs per ton of emissions reduced than those reported here.[18]

---

[17]Some of these increases will undoubtedly be in areas with good air quality and thus will generate little benefit.

[18]An increase in vehicle miles traveled can also cause an increase in congestion. Parry and Small (2002) found the costs of increased congestion to be larger than the costs of increased pollution, although their analysis did not address California or the South Coast Air Basin in particular.

Emission Reductions in All Areas Rather Than Only Where There Are Air Quality Problems. The ZEV program applies throughout California and will result in ZEVs and PZEVs in areas of the state that have no air quality problems. Emission reductions in these areas are of little benefit. Further analysis is needed to determine the importance of this effect in the cost-effectiveness calculations, but any effect will tend to increase cost per ton if "ton" is redefined to restrict emission reductions in areas that need reductions.

Offsetting Interaction with Other Programs. The ZEV program interacts with a number of other regulations that affect its overall impact on emissions. Examples include California's fleet-average NMOG requirement and the federal Corporate Average Fuel Economy (CAFE) standards.

CARB requires that new vehicles sold by each manufacturer meet a fleet-average NMOG requirement. Each different exhaust emission category (the SULEV exhaust standard is one such category) is assigned an NMOG factor, and the average across all vehicles sold by a manufacturer during a year is its fleet-average NMOG. ZEVs, ATPZEVs, and PZEVs count in the fleet-average NMOG calculation, so the ZEV program allows vehicles that do not generate ZEV credits to be dirtier than they otherwise would be. This feedback offsets some of the benefits of the vehicles used to satisfy ZEV program requirements. Accounting for such a feedback will increase the costs per ton above those reported here. It should be noted, however, that this aspect of California's ZEV program could be changed.

The CAFE regulations allow an analogous feedback. Alternative fuel vehicles are granted generous CAFE credits. Thus, the ZEV program allows the fuel efficiency of gasoline vehicles to decline, increasing indirect emissions of NMOG and NOx, as well as of carbon dioxide.[19] The decrease in fuel economy may also offset some of the potential benefit the ZEV program has for reducing dependence on foreign oil. Again, this regulatory provision could be changed. In fact, a recent National Research Council study (2002) recommended that CAFE credits for alternative fuel vehicles be eliminated.

---

[19]As discussed in Subsection 5.1, indirect emissions are negatively related to fuel economy, whereas exhaust emissions and evaporative emissions are not. Exhaust emissions are not related to fuel economy because the emission standards are set in grams per mile. Evaporative emissions are not related to fuel economy because standards for all but running emissions are specified as emissions per test procedure, and standards for running emissions are specified in grams per mile.

7. CONCLUSIONS AND POLICY RECOMMENDATIONS

Even though California has made much progress in improving its air quality over the last 30 years, many parts of the state still violate federal ozone standards. The South Coast Air Basin is far short of the emission reductions required to meet these standards. This shortfall motivates CARB's goal of a zero emission fleet and its first step toward this goal, the ZEV program.

This report examines the costs of and the emission reductions from the various vehicle technologies that manufacturers may use to meet ZEV program requirements. In this concluding section, we summarize our key findings, discuss the conclusions we draw from them, and arrive at implications for ZEV policy.

7.1 COST-EFFECTIVENESS OF TECHNOLOGIES FOR MEETING ZEV PROGRAM REQUIREMENTS

We examined the cost of requiring vehicles to meet progressively tighter emission standards per ton of emissions reduced. We restricted our attention to reductions in non-methane organic gas (NMOG) and oxides of nitrogen (NOx) emissions, which are the key pollutants that must be reduced if California is to meet air quality standards. We examined the cost per ton of moving from CARB's tightest standards outside the ZEV program (the super low emission vehicle [SULEV] exhaust standard and the near-zero evaporative emission standard) to partial zero emission vehicles (PZEVs) and then from PZEVs to ZEVs. We also examined the cost per ton of moving from PZEVs to gasoline electric hybrid vehicles (GHEVs), vehicles that must meet PZEV exhaust and evaporative emission standards but that produce fewer indirect emissions (from fuel extraction, refining, and distribution) than PZEVs do because of their greater fuel efficiency.

We examined costs during the first five years of the program (2003 through 2007) and when the vehicles are in high-volume production. The high-volume cost estimates are based on designs that appear feasible given what is known about battery and fuel-cell technology. They also take into account the types of production processes that are feasible at high volume. Forecasting technological advances is difficult and becomes more so the farther one looks into the future. We thus think it appropriate to interpret our high-volume estimates as the lowest level to which costs are expected to fall over the next 10 years or so given what is currently known about advanced vehicle technologies and manufacturing processes. As manufacturers gain production experience, costs may drop beyond our volume production predictions, but we do not expect these effects to be large over the next 10 years absent significant technological advances. We have two reasons for thinking this way: Our estimates of high-volume costs, particularly our estimates at the low end of the predicted range, assume mature manufacturing techniques; and unexpected hitches may arise in the production process.

Partial Zero Emission Vehicles

Our analysis suggests that PZEVs are an economical way to reduce NMOG and NOx emissions for passenger cars and the smaller category of light-duty trucks (LDT1). We estimate that the cost of reducing emissions from SULEV and near-zero evaporative emission levels to PZEV levels ranges from $18,000 to $71,000 per ton of NMOG plus NOx. The lower end of this range is less than the cost per ton of other regulations that have recently been adopted. The upper end exceeds the costs of recent regulations (all of which are less than $35,000 per ton), but it is plausible that cost per ton will have to rise this high if air quality standards are to be met.

Gasoline Hybrid Electric Vehicles

Whether GHEVs are a cost-effective way to reduce NMOG plus NOx emissions below standard PZEV levels remains uncertain. The outcome rests largely on whether the battery in the hybrid lasts the life of the vehicle. If the hybrid's maintenance costs are comparable to those of a PZEV, then GHEVs will be attractive. However, if the battery needs to be replaced during the vehicle's life, cost per ton may go as high as $180,000. For GHEVs to be cost-effective, the difference between the lifecycle costs of PZEVs and GHEVs must be small, because the difference in NMOG plus NOx emissions between the two vehicles is so small. Because GHEVs have higher fuel economy, their carbon dioxide emissions are lower than those of nonhybrids. GHEVs may thus make sense for reducing carbon dioxide emissions, but an analysis of whether they are the most cost-effective approach was beyond the scope of our study. Electric power usage in conventional vehicles (for electric suspensions, electric steering, etc.) is expected to rise in the future, which means that GHEVs, which already have the capacity in place to accommodate such demand, may become increasingly attractive.

Battery-Powered Electric Vehicles

The cost of reducing emissions from PZEV levels to zero with BPEVs that use nickel metal hydride (NiMH) batteries is high. Vehicles with a 100-mile range most likely require NiMH batteries.[1] We found that the cost per ton of emissions reduced is exceedingly high for full-function EVs using NiMH batteries between 2003 and 2007. Even in volume production, the cost per ton of NMOG plus NOx reduced ranges from $260,000 to $710,000. Both the high costs

---

[1]A 100-mile range is also possible with lithium ion batteries, but this technology has shortcomings of its own (see Anderman, Kalhammer, and MacArthur, 2000).

of full-function EVs and the small difference between the lifetime emissions of PZEVs and ZEVs drive the high costs per ton.

The situation is somewhat better for city EVs with NiMH batteries, but cost per ton still ranges from $70,000 to $370,000 in volume production. On a cost-per-ton basis, city EVs with NiMH batteries thus do not appear to be a very attractive way to reduce emissions. What is more, city EVs are very different from vehicles on the road in California today. This raises questions about whether manufacturers will be able to sell substantial quantities at prices that cover their costs and whether the miles driven in city EVs will fully displace miles driven in conventional vehicles.

We also examined costs for full-function and city EVs powered with lead-acid (PbA) batteries. At the low end of our estimated range, which is based on cost and performance parameters that are plausible but optimistic, city EVs are attractive, and the cost per ton for full-function EVs is arguably not above what might be necessary to meet air quality standards. However, the case for EVs with PbA batteries is still not strong. First, the size and weight of these batteries make it difficult to produce vehicles with broad appeal. Second, the cost per ton of emissions reduced may just as likely turn out to be very high. Our cost estimates top out at $470,000 and $250,000 per ton for full-function and city EVs with PbA batteries, respectively, once transition costs are included (see Table 6.3).

Our analysis leads us to conclude that BPEVs do not appear to be an economical way to reduce NMOG and NOx emissions from PZEV levels. The main stumbling block is still battery cost and energy density, and after a decade of intense development, battery cost and performance remain well short of what is needed for a cost-effective vehicle. CARB's Battery Technical Advisory Panel concluded that major advances or breakthroughs that could reduce battery costs are unlikely through 2006 to 2008. Manufacturing costs could fall below our high-volume estimates as manufacturers gain production experience, but our analysis suggests that substantial reductions are only possible with material cost (battery design) breakthroughs.

Direct Hydrogen Fuel Cell Vehicles

DHFCVs show much more promise than BPEVs, but a great deal of uncertainty remains. Projections that fuel-cell system costs (excluding the fuel tank) can fall to $35 to $60 per kilowatt in volume production are well grounded. Costs might be even lower if the designs with very low platinum loadings pan out, but it is too early to tell if this will be the case. Manufacturing experience may also reduce costs, although we conclude that these reductions may not be large given that most nonmaterial costs have been squeezed out of the low end of our predicted range by mature manufacturing techniques.

If fuel-cell costs do eventually fall to the lower end of the range ($35 per kilowatt), DHFCVs will be an attractive part of California's strategy for meeting ozone standards. However, there is little margin for error. A decline in costs to $60 per kilowatt would be a great technological achievement given where costs are now, but it would not be enough to reduce costs per ton to levels that might be required to meet air quality standards: The cost of reducing NMOG and NOx emissions from PZEV levels is $270,000 per ton at the upper end of our predicted range in volume production.

Much needs to be done to confirm the promise of DHFCVs. The performance and durability of designs with low precious metal loadings need to be tested in the real world. Given recent announcements by several automakers, it appears this testing will begin soon. The automated production processes assumed in volume production need to be validated. And importantly, a fueling infrastructure must be developed. Fuel-cell vehicles face a chicken-and-egg problem regarding fueling infrastructure: A sparse infrastructure limits the attractiveness of DHFCVs, and the low number of DHFCVs limits the number of commercially viable fueling stations. We conclude from our analysis that it is too early to tell whether DHFCVs are an economical way to reduce emissions from PZEV levels to zero.

7.2 ALTERNATIVES FOR MEETING AIR QUALITY STANDARDS

Even if the cost per ton of emissions reduced is high, ZEVs may still make sense as part of California's ozone-reduction strategy if the alternatives for reducing emissions look even less promising. We did not develop a detailed plan for reducing NMOG and NOx emissions to the levels needed to meet ozone standards in the South Coast Air Basin. However, we note that a number of alternatives to ZEVs appear to be available. First, as discussed above, our analysis suggests that PZEVs are cost-effective, and moving the fleet to PZEVs would substantially reduce light-duty vehicle (LDV) emissions. For example, we estimate that if the entire LDV fleet were PZEVs in 2010, fleet NMOG and NOx emissions would be, respectively, 22 and 15 tons per day—levels substantially below the targets of 55 and 103 tons per day called for by the current emission control plan in the South Coast (see Tables 2.1 and 2.2).

Our analysis also shows that substantial emissions are still being contributed by sources other than LDVs. These large sources provide an opportunity to reduce emissions more cheaply than requiring ZEVs. Current emission reduction targets in the South Coast leave 113 tons per day in NMOG emissions from solvent uses other than consumer products and architectural coatings. Nearly 78 tons per day in NMOG emissions also remain in 2010 in the "other stationary source" category.[2] Recent regulations on consumer products are less than $7,000 per ton, and recent regulations on solvents are less than $25,000 per ton. For NOx, the 122 tons per day that remain in 2010 from heavy-duty vehicles and the 94 tons from aircraft are noteworthy. These emissions are much larger than the 25 tons per day of NMOG and 15 tons per day of NOx that would be eliminated by moving from PZEVs to ZEVs. What is more, the costs of recent regulations imposed on these sources have been moderate: less than $7,000 per ton on consumer products, $25,000 on solvents, and $800 on NOx emissions from diesel engines.[3] Strategies to achieve air quality standards at least cost must explore possibilities for reducing emissions from these sources before requiring ZEVs.

The low level of emissions reached when the LDV fleet meets PZEV standards illustrates the small size of the gains to be made from moving the fleet to ZEVs. The low level also brings into question CARB's assumption that a zero emission fleet is necessary to achieve public health goals. While it is true that LDV fleet emissions must be very low relative to historical levels to meet air quality goals, they do not necessarily have to be zero.

7.3 RECOMMENDATIONS FOR ZEV POLICY

We conclude by offering recommendations for California's policy on ZEVs. These recommendations are based on the overall social costs and benefits of the ZEV program. Some program costs might be spread outside California and thus may be of less concern to California policymakers. However, the interests of society as a whole are served by considering costs and benefits whether or not they occur in California. We chose to take this broader view.

Recommendation 1. Drop the Goal of Reducing Fleet Emissions to Zero

CARB should drop its goal of reducing emissions from the vehicle fleet to zero. These emissions do not have to be zero to meet air quality standards in California. Tremendous progress has been made in reducing emissions from internal combustion engine vehicles (ICEVs) since the ZEV program was adopted in 1990, and ZEVs are not necessary for the foreseeable future. Rather than setting a goal of zero emissions for some sources, emission reduction targets should be based on the most cost-effective strategy for reducing emissions from all sources.

---

[2]This category includes waste burning, pesticide application, and farming operations (see Table 2.1).

[3]The South Coast Air Basin is short on NMOG reductions, not NOx reductions. But, as discussed in Subsection 2.1, it may be possible to substitute additional reductions in NOx emissions for NMOG reductions.

Recommendation 2. Require Passenger Cars and Smaller Light-Duty Trucks to Meet PZEV Emission Standards

CARB should gradually require passenger cars and small light-duty trucks (CARB's LDT1 category—trucks with loaded vehicle weight less than or equal to 3,750 pounds) to meet PZEV emission standards. This might be done by gradually reducing the fleet-average NMOG exhaust requirement to PZEV levels for these vehicles and requiring them to meet the zero evaporative emission standard. CARB should factor indirect emissions (those from fuel extraction, processing, and distribution) into the fleet-average NMOG requirement. This assures that vehicles such as gasoline hybrids, which may turn out to be an attractive way to reduce NMOG and NOx emissions, will be appropriately encouraged. CARB should also explore the costs of reducing the emissions of heavier light-duty trucks (the LDT2 category) to PZEV levels.

Recommendation 3. Eliminate the ZEV Requirement

We think a strong case can be made for eliminating the requirement that manufacturers produce ZEVs. We found that ZEVs do not look attractive in terms of cost-effectiveness, but we also looked at some other factors—ones not easy to quantify—that should enter into any assessment of whether to keep or eliminate the ZEV program.

Technology Development. One reason to keep the ZEV requirement is that it may force technology development that would not otherwise occur. Fuel-cell programs have developed substantial momentum, but the fact is that we simply do not know what would happen to fuel-cell research and development if the ZEV requirement were scrapped.[4] On the other hand, keeping the ZEV requirement may push the wrong technology. BPEVs have turned out to be a bad choice, and DHFCVs may turn out to be the same. For example, hydrogen infrastructure problems might prove so significant that fuel-cell vehicles with on-board reformers that generate hydrogen from gasoline or methanol (along with some emissions) might make sense. It may turn out that vehicles that run on compressed natural gas or further-refined gasoline ICEVs are more cost-effective ways to reduce emissions.

Insurance Against Disappointments in Emission Control Technology for Gasoline Vehicles. The ZEV requirement provides some insurance against the possibility that the lifetime emission profiles of PZEVs and other very clean gasoline vehicles will not turn out to be as low as currently projected. PZEVs have not yet been sold, and it will be many years before their in-use emissions can be verified. It may turn out that on-board diagnostics (OBD) systems that

---

[4]Automakers have invested large amounts in fuel-cell technology, and programs such as the California Fuel Cell Partnership and the recently announced federal government-industry fuel-cell partnership (Jones and O'Dell, 2002) suggest that the current pace of fuel-cell development may continue without the ZEV program.

monitor emission performance do not work well when emissions are at such low levels or that emission control systems deteriorate faster than expected. The ZEV program does provide some insurance, but the question is, at what cost. Other types of insurance may be more cost-effective—e.g., research on how to further reduce emissions from stationary sources and from off-road and diesel vehicles.

Reductions in Carbon Dioxide Emissions and Dependence on Foreign Oil. Even if ZEVs do not look like a cost-effective way to reduce NMOG and NOx emissions, the ZEV requirement pushes the development of technologies that can reduce carbon dioxide emissions and dependence on foreign oil. What is uncertain is the value of the ZEV requirement—there may well be much more cost-effective ways to reduce carbon dioxide emissions and dependence on foreign oil. For example, to reduce carbon dioxide emissions, it may be more cost-effective to raise the Corporate Average Fuel Economy (CAFE) standard than to require ZEVs. To reduce dependence on foreign oil, it may make more sense to focus on alternative-fuel vehicles that produce some emissions (such as vehicles that run on compressed natural gas) rather than to require ZEVs.

Some counter that while ZEVs may not be the best way to reduce carbon dioxide emissions or dependence on foreign oil, political necessity argues for the ZEV program. The federal government retains authority to set the CAFE standard, but efforts to increase it have recently been blocked. The argument is that, as the fifth largest economy in the world, California should do whatever it can to reduce carbon dioxide emissions and dependence on foreign oil. The benefit of pushing ahead with ZEVs, however, must be weighed against the costs of adopting a strategy that is inferior to available alternatives.

Uncounted Potential Costs. Our estimates of cost-effectiveness do not include some potential costs. First, the ZEV program may cause the prices of new vehicles to increase and thus slow the sales of new vehicles, thereby causing the average age and emissions of the vehicle fleet to increase. While this feedback is possible in principle, we found that there is a great deal of uncertainty about its size. Second, reductions in operating and maintenance costs may cause drivers to travel more, offsetting the emission benefits of some vehicles. Given the ZEV's range and fueling infrastructure limitations, however, it seems unlikely that ZEVs will encourage more travel. Finally, interactions with CARB's fleet-average NMOG requirement and the federal CAFE standard will likely offset some of the ZEV emission reductions. ZEVs are included in the calculation of fleet-average NMOG emissions and fuel economy and thus allow the non-ZEV portion of the fleet to have higher emissions or lower fuel economy than it would otherwise. In principle, these regulations can be rewritten, but there are often political obstacles to doing so. It is difficult to assess the overall importance of these uncounted potential costs, but we know they work against any uncounted potential benefits.

Synthesis. There is room for disagreement over whether the difficult-to-quantify benefits outweigh the difficult-to-quantify costs, but we conclude, overall, that the ZEV program should be ended. What concerns us most about the ZEV requirement is that it focuses on a very narrow set of technologies in its aim to reduce air pollution in California. The focus on zero emission technologies seems particularly inappropriate given that ZEVs are not needed to meet air quality standards and that alternatives for reducing emissions at lower costs appear to be available. Arguments for ZEVs that are based on reducing carbon dioxide emissions or dependence on foreign oil are also not convincing. These two social goals may well be important, but much more needs to be done to show that ZEVs are a cost-effective way to achieve them.

Recommendation 4. If the ZEV Requirement Cannot Be Eliminated, Delay It or Reduce the Number of Fuel-Cell Vehicles Needed to Satisfy It in the Early Years

As it stands now, the ZEV program will force at least some manufacturers to produce expensive BPEVs during its initial years. BPEVs show little long-term promise, and it does not make sense to continue investing in this technology. By delaying the program, CARB would allow time to evaluate the promise of DHFCVs, the only other zero emission technology that appears viable for the foreseeable future.

If CARB does not want to delay the program because it believes doing so will substantially reduce the incentives to bring the potentially promising direct hydrogen fuel-cell technology to the market, it might, as an alternative, significantly reduce the number of DHFCVs needed to satisfy the ZEV requirement. This could be done by increasing the number of ZEV credits generated by each DHFCV. A substantial increase would prevent a large number of battery-powered or fuel-cell vehicles from being put on the road but would still allow CARB and the manufacturers to better understand the real-world potential of fuel-cell vehicles.

Whether the program is delayed or the required number of DHFCVs is reduced by increasing the credit multiplier, CARB should periodically assess the progress being made in DHFCVs. If it becomes clear that DHFCVs are not a cost-effective way to reduce emissions, the ZEV program should be abandoned. If DHFCVs live up to their potential, the credit multiplier could be gradually reduced and production increased.

Recommendation 5. Focus on Performance Requirements, Not Technology Mandates

CARB should focus on setting performance requirements and let the automakers determine how best to achieve them. It should continue to set very stringent fleet-average emission requirements for the new vehicle fleet, but there is no need to set allowable emissions to zero or to require manufacturers to meet the average in part with ZEVs.

California has made remarkable progress in improving its air quality over the last 30 years. However, much remains to be done. Policymakers should seek to facilitate and encourage technological transformation while allowing the flexibility needed for different technologies to compete. Allowing such flexibility would be a departure from the ZEV program, but one worth pursuing.

APPENDIX

A. CHARACTERISTICS OF BATTERY-POWERED ELECTRIC VEHICLES PRODUCED TO DATE

A.1 FULL-FUNCTION ELECTRIC VEHICLES

Table A.1 lists the performance characteristics of the full-function EVs that large-volume manufactures have produced to date. These vehicles were sold or leased between 1997 and 2000 to satisfy the memoranda of understanding between CARB and the automakers. For each full-function EV, the table reports the performance characteristics of an internal combustion engine vehicle (ICEV) that is comparable in size and body style.

EV range on a single charge generally varies from 50 to 100 miles, with the exception of GM's NiMH-powered EV1. Where comparisons are available, the EVs have substantially lower top speeds than the comparable ICEV and, again with the exception of GM's EV1, accelerate more slowly.

A.2 CITY ELECTRIC VEHICLES

City EVs offer lower performance than full-function EVs, with concomitantly less demand on the batteries. They are ultracompact (105 inches to 120 inches long), two-passenger cars with top speeds and acceleration that make them fit primarily for surface road use. They are intended, ultimately, to meet all Federal Motor Vehicle Safety Standards (FMVSS) for passenger cars, including, for example, having dual air bags.[1] Four of the auto manufacturers subject to the ZEV program have announced their intention to introduce city EVs.

In January 1999, Ford acquired the bankrupt Norwegian company Pivco, manufacturer of the Citybee city EV.[2] The Citybee and its successor, the Th!nk City,[3] have had Scandinavian sales of over 500 (*EV World*, 2000) at a price of approximately $25,000 (USA Today, 2000). The City has a top speed of 56 mph, a 0-30 mph acceleration of 7 seconds (*Automotive Intelligence*

---

[1]No currently available city EVs are FMVSS 591 certified. Certification is required only for production models with U.S. sales of more than 2,000.

[2]Forty Citybees were used in a 1995 station car demonstration in the San Francisco area (Moore, 2000a).

[3]The Th!nk Group is intended to become a separate brand within the Ford Motor Company, home for all its alternative-power vehicle technologies (New Car Test Drive, 2001). Th!nk Mobility, LLC, is a wholly owned subsidiary dedicated to BPEVs, and Th!nk Technologies is dedicated to fuel-cell EVs (Burke, Kurani, and Kenney, 2000).

Table A.1

Full-Function Electric Vehicles Leased in California Between 1997 and 2000

Full-Function EV Make and Model	Battery Type	Markets Where Available[a]	0-50 mph (seconds)	Top Speed (mph)	Range (miles)[b]		Comparable ICEV		
					City[c]	Freeway[d]	Make and Model	0-60 mph (seconds)	Top Speed (mph)
Chevy S-10	PbA	F	10.9	71	41/50	48/57	S-10	8.8	n/a
Chevy S-10	NiMH	F	10.4	71	63/70	80/84	S-10	8.8	n/a
DC EPIC	NiMH	F	11.2	78	68/82	75/99	Plymouth Voyager	12.1	n/a
Ford Ranger	PbA	R,F	12.3	75	60/72	57/66	Ranger	8.8	n/a
Ford Ranger	NiMH	R,F	11.2	75	73/81	71/76	Ranger	8.8	n/a
GM EV1 (Gen I)	PbA	R,F	6.7	85	65/80	78/91	Mazda Miata	8.8	113
GM EV1 (Gen II)	NiMH	R,F	6.5	80	140[e]		Mazda Miata	8.8	113
Honda EV+	NiMH	R,F	17.7[f]	85	87/105	79/89	Honda CRV	11.4	11.4
Nissan Altra	LiIon	F	12.0	75	95/122	82/94	Saturn SW2	8.1	130
Toyota RAV4	NiMH	F	12.9	78	84/93	78/82	RAV4	10.1	99

[a]F=fleets, R=retail customers.

[b]Southern California Edison (SCE) Performance Characterization Summary.

[c]Urban range with accessories/Urban range without accessories (UR2/UR1).

[d]Freeway range with accessories/Freeway range without accessories (FW2/FW1).

[e]SCE data unavailable. Range for drive cycle test reported in "Baseline Performance (EVAmerica) Testing Trends" test.

[f]0–60 miles per hour.

News, 2000), and a range of approximately 60 miles (Ford Motor Company, 1999) on a 114-volt NiCd battery pack (Brown, 2001).

Ford has a manufacturing capacity of 5,000 City vehicles per year at its Th!nk Nordic AS plant (Walsh, 1999), and a domestic manufacturing site is possible if demand warrants it (Wallace, 2000). In 2000, Ford announced plans for 47 dealers that already sell the Ranger EV to sell the City by the fourth quarter of 2001 (Connelly, 2000). No price was given, but a Ford executive reported telling Pivco that the target price was $15,000 (Wallace, 2000), and a Pivco official earlier suggested an expected U.S. price of $20,000 (Moore, 2000a). Following on a 2000 U.S. demonstration program of about 700 vehicles (300 in California) (CARB, 2000b), the City became available in 2001, for 34-month lease only, at $200 per month (Brown, 2001).[4] The City will be offered for sale in 2002, pending FMVSS certification.[5]

The Nissan Hypermini has a top speed of 60 mph and a range of 80 miles on lithium ion (LiIon) batteries (Tribdino, 2001).[6] The Hypermini is currently for sale in Japan for approximately $35,000 (Nissan, 2000); 120 vehicles have been sold to city officials and private citizens in car-sharing programs in three cities. No plans have been announced for U.S. sales, but at least 30 vehicles have been sent to the United States for trials (Moore, 2000b), including those at the Los Angeles Department of Water and Power (Brungard, 2001) and at UC Davis (2001).

Toyota and Honda have also developed city EVs. The Toyota e_com has a top speed of 62 mph and a range of 60 miles on NiMH batteries (Electric Vehicle Association of America [EVAA], 2001); 12 vehicles are currently being tested at UC Irvine (2001). The Honda City Pal has a top speed of 68 mph and a range of 80 miles on NiMH batteries (EVAA, 2001); it is currently available for rent on a trial basis in Singapore (*Singapore Straits Times*, 2001).

---

[4]Because the current models are not FMVSS certified, they may stay in the United States for only 36 months.

[5]A notice from Ford of Norway to Norwegian Th!nk City owners says that a new model, meeting all U.S. safety standards and with a higher top speed of 65 mph (at the expense of some range), will be offered in the United States in January 2002. Translation at groups.yahoo.com/group/think_ev/message/222 (viewed November 5, 2001).

[6]A more recent press release claims a "real-world" driving range of 30-35 mph (Nissan, 2001).

B. ASSUMPTIONS USED TO CALCULATE VEHICLE PRODUCTION VOLUMES

This appendix details the assumptions used to calculate the production volumes for the scenarios in Subsection 3.2.

The number of ZEV credits generated by a ZEV is the product of

- the ZEV phase-in multiplier
- the extended range multiplier
- the high-efficiency multiplier.

Table B.1 shows the assumptions we used to calculate the generated credits. The first row shows the phase-in multiplier specified by CARB. The extended range multiplier is calculated using the following: (urban all-electric range − 25)/25. For the full-function EV, we assume a vehicle with an urban all-electric range of 125 miles, which roughly corresponds to a real-world range of 100 miles. For the city EV, we assume an electric all-urban range of 55 miles. The resulting range multipliers are listed in the second row of Table B.1.

The high-efficiency multiplier is

- CMPEG/(1.5*Baseline Fuel Economy) from 2002 to 2007
- CMPEG/(2.0*Baseline Fuel Economy) from 2008 on

where CMPEG is the California miles per equivalent gallon and baseline fuel economy is the gas mileage for the reference vehicle. We base our high-efficiency multiplier for the full-function EV on data from the RAV4. For the city EV, we use data from Toyota's e_com (as reported in CARB, 2001e, p.19). The parameter values and resulting high-efficiency multipliers are listed in the bottom group of rows in Table B.1.

The range multiplier is phased out over time, and the high-efficiency multiplier is phased in. The phased multipliers used to calculate the overall ZEV credit are determined by

- ((range multiplier − 1)*range phasing factor) + 1
- ((efficiency multiplier − 1)*efficiency phasing factor) + 1

for the range and efficiency multipliers, respectively. The phasing factors are given in CARB, 2002 (p. 16). (The weight on the range multiplier declines from 1.0 in 2004 to 0.15 in 2012.)

Table B.2 shows the resulting credit multipliers for full-function and city EVs when the phase-in, range, and high-efficiency multipliers are combined.

The number of credits for an advanced technology partial zero emission vehicle (ATPZEV) is

- (0.2 + advanced ZEV componentry allowance)*(PZEV phase-in multiplier)*(ATPZEV high efficiency multiplier).

Table B.1

Assumptions Used to Calculate Credits Generated by Full-Function EVs and City EVs

Multiplier	Full-Function EVs	City EVs
ZEV phase-in multiplier	1.25 from 2003 to 2005; 1 from 2006 on	1.25 from 2003 to 2005; 1 from 2006 on
Extended range multiplier	4	1.2
High-efficiency multiplier		
Baseline fuel economy	25.0	30.6 from 2003 to 2007; 45.9 from 2008 on
CMPEG[a]	102.6	127.4
Multiplier	2.74 from 2003 to 2007; 2.05 from 2008 on	2.78 from 2003 to 2007; 1.38 from 2008 on

[a]California miles per equivalent gallon.

Table B.2

Number of ZEV Credits Per Vehicle

Year	Full-Function EV	City EV	ATPZEV
2003	5.00	1.50	2.17
2004	5.00	1.50	1.08
2005	5.10	1.72	0.70
2006	4.51	1.82	0.54
2007	4.60	2.16	0.54
2008	3.10	1.30	0.54
2009	3.04	1.33	0.54
2010	2.94	1.36	0.54
2011	2.94	1.36	0.54
2012	2.71	1.35	0.54
2013	2.71	1.35	0.54
2014	2.71	1.35	0.54
2015	2.71	1.35	0.54
2016	2.71	1.35	0.54
2017	2.71	1.35	0.54
2018	2.71	1.35	0.54
2019	2.71	1.35	0.54
2020	2.71	1.35	0.54
2021	2.71	1.35	0.54
2022	2.71	1.35	0.54
2023	2.71	1.35	0.54
2024	2.71	1.35	0.54
2025	2.71	1.35	0.54
2026	2.71	1.35	0.54
2027	2.71	1.35	0.54
2028	2.71	1.35	0.54
2029	2.71	1.35	0.54
2030	2.71	1.35	0.54

The ATPZEV phase-in multiplier is specified in CARB, 2002 (p. 9). To calculate the high-efficiency multiplier, we use the formula above and data from CARB, 2001e (p. 19) on the Toyota Prius gasoline hybrid electric vehicle (GHEV). The Prius has a CMPEG of 57.7 miles with a baseline fuel economy of 30.4.

The advanced componentry allowance is based on the reduction of carbon dioxide emissions using the method specified by CARB (2002, p. 8). The result for a Prius is 0.23. The resulting number of credits per GHEV is listed in the last column of Table B.2.

C. COMPONENTS OF INCREMENTAL COST IN VOLUME PRODUCTION AND SENSITIVITY ANALYSIS

C.1 COMPONENTS OF HIGH-VOLUME INCREMENTAL COST

Tables C.1 through C.4 break down our estimated incremental vehicle cost in volume production into several categories. Separate cost breakdowns are reported for battery-powered electric vehicles (BPEVs) that use NiMH and PbA batteries.

For full-function EVs, we assume a pack size of 30-kWh for the high end of our estimates and 24-kWh for the low end. The smaller pack size reflects the possibility of increased vehicle efficiency (see Subsection 4.7). For city EVs, we set pack size to 10 and 8 kWh for the high and low cost estimates, respectively. The multipliers used to calculate the change in indirect costs are as follows: 0.15 for the low estimate for battery packs (modules and auxiliaries) and hydrogen fuel-cell systems (excluding fuel tank), 0.30 for the high estimate of battery packs and fuel-cell systems, 0.5 for other electric components and the hydrogen fuel tank, and 1.0 for the internal combustion engine (ICE).

A direct hydrogen fuel-cell vehicle (DHFCV) with a 60-kW proton exchange membrane (PEM) and a 25-kW battery costs more than a DHFCV with a 25-kW PEM and a 60-kW battery. We assume that manufacturers will select the configuration that is least costly, so use the 25-kW PEM and 60-kW battery configuration in our analysis of volume production costs.

C.2 SENSITIVITY OF INCREMENTAL COST TO VEHICLE SPECIFICATIONS

Here we consider how incremental cost changes as a result of changes in vehicle specifications. CARB (2000b) provides ranges for full-function EV and city EV characteristics; those for energy storage and peak power are quite large (see Table C.5). The wide range reflects the range of vehicles that have been produced. For example, Toyota's RAV4 has a 50-kW peak motor whereas an EV1 has a 100+ kW motor.

Figures C.1 and C.2 show how the average incremental cost for a full-function EV between 2003 and 2010 varies with motor peak power and battery size. Clearly, as power and range vary, so does the cost increment relative to an internal combustion engine vehicle (ICEV). Nonetheless, the eight-year (2003-2010) average cost increment for a full-function EV with an NiMH pack is never below $9,800, given the ranges from the table.

Table C.1

Incremental High-Volume Production Cost Breakdown for Full-Function EVs

($)

Component	NiMH Batteries		PbA Batteries	
	Low Estimate	High Estimate	Low Estimate	High Estimate
Battery modules	5,400	7,500	2,400	3,000
Battery auxiliaries	192	240	336	420
Motor and controller	725	1,200	725	1,200
Conductive charger	250	300	250	300
Other (transmission, electric steering, etc.)	168	210	168	210
ICE savings (85-kW engine)	-1,530	-1,360	-1,530	-1,360
Subtotal	5,205	8,090	2,349	3,770
Indirect costs	-120	1,817	-548	521
Total	5,085	9,907	1,801	4,291

Table C.2

Incremental High-Volume Production Cost Breakdown for City EVs

($)

Component	NiMH Batteries		PbA Batteries	
	Low Estimate	High Estimate	Low Estimate	High Estimate
Battery modules	1,800	2,500	800	1,000
Battery auxiliaries	64	80	112	140
Motor and controller	450	550	450	550
Conductive charger	250	300	250	300
Other (transmission, electric steering, etc.)	56	70	56	70
ICE savings (85-kW engine)	-765	-680	-765	-680
Subtotal	1,855	2,820	903	1,380
Indirect costs	-107	554	-250	122
Total	1,748	3,374	653	1,502

Table C.3

Incremental High-Volume Production Cost Breakdown for GHEVs

($)

Component	Low Estimate	High Estimate
20-25kW NiMH battery	500	500
Motor and controller	750	851
ICE savings (assumes motor is 33-kW smaller than in non-hybrid vehicle)	-594	-528
Subtotal	656	823
Indirect costs	-144	47
Total	512	870

Table C.4

Incremental High-Volume Production Cost Breakdown for DHFCVs

($)

Component	25-kW PEM + 60-kW Battery		60-kW PEM + 25-kW Battery	
	Low Estimate	High Estimate	Low Estimate	High Estimate
60-kW PEM system	875	1,500	2,100	3,600
Motor and controller	725	1,200	1,100	1,800
Hydrogen tank	336	652	336	652
Energy storage (25-kW NiMH battery)	1,200	1,200	500	500
Other[a]	168	210	168	210
ICE savings	-1,530	-1,360	-1,530	-1,360
Subtotal	1,774	3,402	2,674	5,402
Indirect costs	-604	481	-338	1,201
Total	1,170	3,883	2,336	6,603

[a]Transmission, electric steering, etc.

Table C.5

Range of Vehicle Specifications Cited in CARB Staff Analysis

Vehicle Type	Energy Storage Capacity (kWh)	Drive System Peak Power (kW)	Examples
City EV	10-15	20-30	Toyota e_com, Nissan Hypermini, Th!nk City
Full-function EV	15-35+	50-150	EV1, EV-Plus, RAV4 EV, Altra

SOURCE: CARB, 2000b.

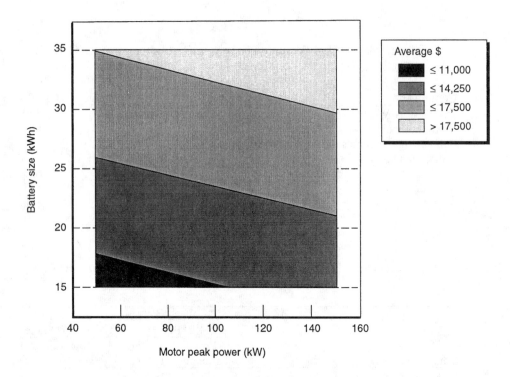

Figure C.1 Average Cost Increment for Full-Function EV with NiMH Batteries Between 2003 and 2010

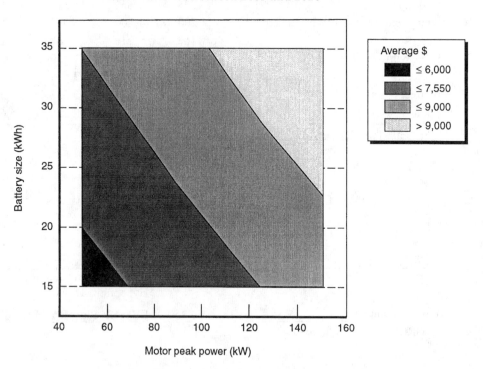

Figure C.2 Average Cost Increment for Full-Function EV with PbA Batteries Between 2003 and 2010

D. METHODS USED TO CALCULATE ELECTRICITY AND HYDROGEN COSTS

This appendix describes the methods for deriving the electricity and hydrogen costs used in our analysis.

D.1 ELECTRICITY COSTS

We assume that electric vehicles (EVs) are charged at off-peak rates with power generated by a modern, efficient, natural gas-fired, baseload power plant. We first discuss the fuel costs of generating electricity. The most efficient large combined-cycle plants require approximately 6,300 BTUs to generate 1 kilowatt-hour (*Gas Turbine World*, 2000).[1] Costs of natural gas delivered to a utility under contract can be expected to vary between $4 and $6 per million BTUs (MMBTU), which implies a fuel cost of from $0.041 to $0.058 per kilowatt-hour.

The construction cost of such a plant runs on the order of $500 per kilowatt. A reasonable estimate of the annual amount needed to pay back construction costs and the cost of capital is 15 to 20 percent of the construction (Bartis, 2001). This results in a capital cost of $75 to $100 per year per kilowatt-hour. If the plant runs 70 percent of the time, this amounts to between $0.012 and $0.016 per kilowatt-hour. We allow the capital cost to vary between $0.012 and $0.016 per kilowatt-hour.

Table D.1 lists ranges for the fuel and capital costs as well as the other components of electricity costs. The transmission and distribution costs and metering costs are taken from Southern California Edison's rates for domestic charging of EVs (Southern California Edison, 2000). Estimates of typical plant operations and maintenance (O&M) costs are from James T. Bartis, a RAND energy expert (Bartis, 2001).

If the number of EVs remains low and they are charged off peak, then it is possible that they could be charged using existing generating capacity and the existing transmission and distribution system. In this case, the social costs of electricity would only be $0.035 to $0.048 per kilowatt-hour. We think this range appropriate for the low volume of vehicles that will be produced between 2003 and 2007. In volume production, the number of EVs on the road will be substantial, in which case additional generation and transmission capacity would have to be built. We thus think it appropriate to use electricity rates that include the full capital and transmission costs for our high-volume lifecycle cost estimates.

---

[1]Gas prices are expressed in terms of higher heating values, whereas power-plant heat rates are expressed in terms of lower heating values. The 5,650-BTU/kWh heat rate listed for this particular power plant was increased by 9.9 percent to convert to the higher heat units used in natural gas prices.

Table D.1

Estimates of Components of Electricity Cost
($ per kWh)

Cost Category	Lower End of Range	Upper End of Range
Fuel	0.025	0.038
Capital	0.012	0.016
Plant O&M	0.005	0.005
Transmission and distribution	0.025	0.025
Metering	0.005	0.005
Total	0.072	0.089

SOURCES: Southern California Edison, 2000, and Bartis, 2001.

Table D.2

Parameters Used to Calculate Costs of Hydrogen

Parameter	2006-2010		High-Volume Production	
	Used for Low-Cost Estimate	Used for High-Cost Estimate	Used for Low-Cost Estimate	Used for High-Cost Estimate
Cost of natural gas ($/MMBTU)	5.5	7.5	5.5	7.5
Heat rate (MMBTU/kg hydrogen)	0.192	0.192	0.192	0.192
Capital cost ($)	450,000	450,000	220,000	220,000
Rate of return per year	0.15	0.20	0.15	0.20
O&M costs ($/fill-up)	0.49	0.49	0.49	0.49
Vehicles per day	60	50	75	75
Hydrogen per fill-up (kg)	4	4	4	4

D.2 HYDROGEN COSTS

We develop estimates of the costs of refueling direct hydrogen fuel-cell vehicles (DHFCVs) at a fueling station, with hydrogen generated on site from natural gas using a reformer. Our estimates are based on an analysis of fueling station costs by Unnasch (2001).

Table D.2 lists the parameter values used in our analysis. Natural gas will cost more to deliver to service stations than to large power plants, so we set the price of natural gas to $5.50 to $7.50 per MMBTU. Unnasch provides data that allow us to infer the amount of natural gas required to generate a kilogram of hydrogen. He estimates that at $5.50 per MMBTU, $1.06 of natural gas will generate a kilogram of hydrogen, which implies a "heat rate" of 192,000 BTUs per kilogram of hydrogen. For 2006 to 2010, we use Bevilacqua-Knight's estimated cost of $450,000 for a natural gas-based reformer and associated vehicle fueling equipment that can serve 50 to 75 vehicles a day (Bevilacqua-Knight, 2001, p. E-4). This cost is supposed to represent the unit cost of the first 500 units installed. The authors note that this cost represents a "fairly aggressive" reduction in hydrogen fueling equipment costs from current levels. To calculate hydrogen fuel costs when DHFCVs reach volume production, we reduce the fueling station costs to $220,000 per fueling dispenser, which is the cost of adding an additional hydrogen dispenser at an existing hydrogen fueling station (Bevilacqua-Knight, 2001, p. E-4). As in our analysis of electricity cost, we allow annual capital costs to vary between 15 and 20 percent of equipment cost.[2]

Unnasch's analysis (2001, pp. E-4, E-5) implies that the O&M costs are $0.49 per fill-up at gasoline stations, and we adopt this figure in our analysis. The hydrogen reformer is sized to service 50 to 75 vehicles a day. Unnasch reports that gasoline stations typically serve 72 vehicles per nozzle per day. We assume that each hydrogen nozzle serves 50 to 60 vehicles a day (60 is used for the low hydrogen cost) in the short run, which increases to 75 a day in the long run. Finally, based on initial specifications of vehicles being developed by Honda, we assume that each vehicle takes 4 kg of hydrogen per fill-up.

The component and total costs of hydrogen are reported in Table D.3. Fuel costs are calculated by multiplying the cost of natural gas by the heat rate. The cost of compressing natural gas is taken from Unnasch, 2001 (p. E-12); we caution, however, that we do not know what electricity rates Unnasch used to calculate compression costs.[3] Finally, O&M costs are calculated by dividing the O&M cost per fill-up by the amount of hydrogen per fill-up. Note that, as before, we do not include taxes in our costs. We find that hydrogen costs fall between $2.32 and $3.16 per kilogram between 2006 and 2010, and between $1.85 and $2.33 once DHFCVs reach volume production.[4]

Table D.3

Estimates of Components of Hydrogen Cost
($ per kg)

Cost Category	2006-2010	High-Volume Production
Capital	[0.77, 1.23]	[0.30, 0.40]
Fuel	[1.06, 1.44]	[1.06, 1.44]
Compression	0.37	0.37
O&M	0.12	0.12
Total	[2.32, 3.16]	[1.85, 2.33]

[2]We calculate capital cost per kilogram of hydrogen by dividing annual capital cost by the amount of hydrogen delivered per year (the number of fill-ups per year times the number of kilograms per fill-up). The number of fill-ups per year is 365 times the number of vehicles per day (see Table D.2 for number of vehicles per day).

[3]A $0.37 compression cost means that between 4 and 5.2-kWh are needed per kilogram of hydrogen when electricity costs $0.07 to $0.09 per kilowatt-hour.

[4]As in the case of EVs, one might argue that hydrogen for fuel-cell vehicles can be produced using excess electrical and natural gas capacity. As before, however, we are interested in a scenario with substantial numbers of fuel-cell vehicles and thus use fully burdened electricity and natural gas costs in our calculations of hydrogen costs.

E. WILLINGNESS TO PAY FOR FULL-FUNCTION AND CITY ELECTRIC VEHICLES

In this appendix, we examine available information on consumer willingness to pay for battery-powered electric vehicles (BPEVs). We first review the information available on full-function EVs and then turn to city EVs.

E.1 WILLINGNESS TO PAY FOR FULL-FUNCTION ELECTRIC VEHICLES

We base our analysis on two sources: the experiences of the large-volume manufacturers to date with full-function EVs and stated-preference studies on the demand for these vehicles.

Manufacturer Experience with Full-Function EVs

Table E.1 lists the numbers of vehicles leased and their lease rates from 1996, when the first full-function EVs were available, through March 2000. It also lists the lease rates for comparable internal combustion engine vehicles (ICEVs). With the exception of the Ford Ranger and Chevy S-10 with PbA batteries, vehicles were available only for lease. Manufacturers chose to lease the vehicles because of uncertainty about battery lifetime, as well as their concern that the vehicle and battery technology would quickly become obsolete. Some of the manufacturers also leased and sold vehicles outside California, although across all manufacturers, at least two-thirds were in California. Here we focus on the experience in California.

Lease rates for full-function EVs were higher than those for comparable ICEVs, except in the case of the EV1 with PbA batteries. These rates were in part due to special features of the leases: leases usually covered all maintenance of the vehicle (in addition to the bumper-to-bumper warranties available on most new cars) and, in some cases, included roadside assistance. Honda's lease rate included collision insurance, which might account for upwards of $70 of the $455 lease rate. The lessee, however, usually had to pay at least part of the charger installation cost (see Subsection 4.4).

The high lease rates indicate that at least some consumers are willing to pay a premium for a full-function EV. However, experience to date does not suggest that the number of consumers willing to pay such a premium is large. The number of vehicles required under the ZEV program is many times the number produced to date. Roughly 2,500 full-function EVs were sold or leased between 1997 and 2000. Even our low-volume scenario for full-function EVs (see Subsection 3.2) shows manufacturers producing 4,000 vehicles a year during the early years of the program, and the number rises to 20,000 by 2012 and nearly 40,000 by 2020 (see Figure 3.1).

Table E.1

Lease Rates of Electric Vehicles Leased in California Between 1997 and March 31, 2000

					Comparable ICEV		
Full-Function EV Make and Model	Battery Type	Market Where Available	Lease Price[a,b]	Number Leased in Calif.[b]	Make and Model	Zero-Down 3-Year Lease Rate	Approxi-mate MSRP[c]
Chevy S-10	PbA	F	439	110	ICEV S-10	225	15,000
Chevy S-10	NiMH	F	440	117	ICEV Gas S-10	225	15,000
DC EPIC	NiMH	F	450	185	Voyager	325	21,000
Ford Ranger	PbA	R,F	349	52	ICEV Ranger	200	12,500
Ford Ranger	NiMH	R,F	450	356	ICEV Ranger	200	12,500
GM EV1 (Gen I)	PbA	R,F	349	400	Mazda Miata	375	26,000
GM EV1 (Gen II)	NiMH	R,F	499	162	Mazda Miata	375	26,000
Honda EV+	NiMH	R,F	455[d]	276	Honda CRV	300	19,000
Nissan Altra	LiIon	R,F	599	81	Saturn SW2	275	18,000
Toyota RAV4	NiMH	F	457	486	ICEV RAV4	300	20,500

[a]Price paid by consumer (net of subsidies).
[b]CARB, 2000b, p. 11.
[c]MSRP = manufacturer's suggested retail price.
[d]Includes collision insurance.

Below we review the leasing experience of each manufacturer to better understand the market for full-function EVs. For the most part, this review shows that manufacturers had difficulty moving the vehicles at the lease rates offered. The descriptions are based on publicly available information, such as automaker presentations at CARB workshops.

DaimlerChrylser EPIC. Chrysler built 207 NiMH EPIC minivans in May 1998, and it took nearly 2.5 years to lease them all. The last one was leased in August 2000. All leases were to fleets.

Ford Ranger. Ford leased NiMH Rangers in California to meet the terms of its 1996 memorandum of agreement (MOA). The leases were priced at $450 a month, and the vehicles moved slowly at this price: Between mid-1998 and mid-1999, Ford leased only 42 of the 205 vehicles required by the MOA. Ford reduced the monthly price to $199 at the end of 1999 and leased the remaining 163 vehicles in a few months. This lease rate is comparable to the lease rate for a gasoline Ranger.

GM EV1. GM offered EV1s with PbA batteries for sale in California and Arizona starting in December 1996. Sales during the first three months were encouraging but soon slowed thereafter. It took 27 months (until February 1998) to sell the first batch of 500 vehicles built. The vehicles were initially offered at $477 a month, but the rate was soon reduced to $399 and then reduced to $349 in mid-1998. The price reductions did little to stimulate sales.

GM was further disappointed by the results of an employee lease program at Southern California Edison. Edison and some automakers joined to subsidize lease rates of their EVs to Edison employees. In the latter part of 1997, GM offered low-mileage demonstration EV1s at $249 per month, but only two vehicles were leased. In 1998, new EV1s were offered at $299 a month, but there were no takers (Stewart, 2000).

GM also produced a batch of second-generation EV1s; these vehicles moved slowly. GM has suspended production of the EV1 and has publicly released no plans to produce more full-function EVs.

Honda EV+. It took Honda longer than it initially expected to lease the 276 EV+'s that were on the road by March 2000. Honda had to expand the number of dealers that offered the vehicle and also had to compromise on its objectives for the split of sales between households and fleets. Originally, Honda hoped to lease 75 percent of its EV+'s to households, a market it viewed as critical to the future commercial viability of its EVs (CARB, 2000b, p. 71). To date, however, about one-half of the vehicles leased in California have gone to households. The EV+ was discontinued in 1999.

Nissan Altra. The lease rate for the Altra was high ($599 per month), but the number of vehicles offered for lease to August 2000 was limited (81). The Altra is not currently manufactured, and no new leases are available.

Toyota RAV4. Toyota began leasing the RAV4 for $457 per month and appears to have had little trouble leasing the vehicles it has produced. It is the only large-volume manufacturer that continues to produce a full-function EV. Toyota initially leased only to the fleet market. When it tried to lease to consumers, the results were not encouraging. The RAV4 was offered to Edison employees at $250 a month, but there was little response. In late 2001, Toyota announced plans to sell the RAV4 in California beginning in February 2002 (Hart, 2001). The MSRP is $42,000, but state and federal credits will knock the price down to $30,000.

Market Research on Demand for Full-Function EVs

Below we highlight and interpret findings of several stated-preference studies on willingness to pay for full-function EVs. These studies are based on surveys that ask households or businesses how much they are willing to pay for different types of vehicles. Stated-preference models of vehicle choice may exaggerate how much consumers value BPEVs, as survey respondents may want to indicate their support for public policies to reduce pollution when there is no cost to them to do so (i.e., they do not have to actually buy the vehicles they select in a hypothetical situation). Dixon and Garber (1996, pp. 274-286) provide a more detailed review of studies completed by 1996.

Hill. Hill (1987) analyzed data from 474 respondents in a mid-1983 phone survey of managers of commercial vehicle fleets. Each respondent was presented with a series of vehicles characterized by their range and lifecycle cost relative to a conventional vehicle, and was asked whether each vehicle would be useful to the operations of the respondent's fleet and, if so, how many the respondent would buy. The three ranges considered were 30, 60, and 90 miles, and the three lifecycle costs were 10 percent less, 15 percent more, and equal to the lifecycle costs of a conventional vehicle. Econometric analysis of the data led Hill (1987) to conclude that his analysis "provides strong evidence that firms would be willing to cope with the limited range of electric vehicles *if* these vehicles were able to provide a *less* costly means of doing business" (p. 284) [emphasis added].

Turrentine and Kurani. Turrentine and Kurani (1995) interviewed 454 California households on their preferences regarding EVs. Study participants were asked to choose, based on written and video descriptions, between a gasoline vehicle and an EV, each of which had several body styles and options. Turrentine and Kurani predicted the quantities of EVs that would be demanded at prices equal to those of ICEVs comparable in size, body style, styling, and optional features. In their "Choice One Situation," which is most relevant to the types of full-function EVs that manufacturers might offer in 2003, consumers could choose between a battery pack that would provide a range of 80-100 miles and a more expensive pack that would extend the range to 100-120 miles. The EV with the 80-100 mile range had operating costs similar to those of a comparable ICEV.

Turrentine and Kurani conclude that "The results of this study give strong evidence of a market for EVs large enough to fulfill the year 1998 and 2001 mandates with current electric vehicle and battery technologies" (p. 10). (At the time of the Turrentine and Kurani study, the program requirement was 2 percent in 1998 and 5 percent in 2001.) They were overly optimistic about the EV technologies that would be available in 1998 and 2001, but their assumptions seem more reasonable for 2003. We thus interpret their results to indicate that full-function EVs could account for 5 percent of household sales if (a) the prices to households of full-function EVs were equal to those of comparable ICEVs, and (b) the lifetime operating costs of full-function EVs were equal to those of ICEVs.

Turrentine and Kurani provide no direct information for their Choice Situation One on the quantities of EVs demanded at prices above those of comparable ICEVs. However, they note that "Our previous research, though informal, seems to confirm the opinion that not many consumers will pay extra for electric vehicles" (p. 9).

Kavalec. Kavalec (1996) projected the number of EVs that would be bought at different prices. His analysis is based on an extension of the CALCARS model, which is based on the stated vehicle preferences of 4,750 households surveyed in 1993.[1]

Kavalec found that "To achieve the ten percent requirement in 2003, electric vehicles had to be priced $2,000 (in 1993$) less than similar gasoline vehicles. . . . By 2005, the mandate could be met with the same prices for gasoline and electric vehicles" (p. 31). The results are based on a set of vehicle characteristics that seem reasonable in some cases but overly optimistic in others. The assumptions on incremental EV operating costs and the 80-mile range, for example, seem reasonable,[2] but the vehicle efficiency, top speed, service station availability, and recharge time do not.[3] It is hard to understand why the price differential disappears in 2005 when the only apparent difference is that EV range in 2005 is 90 rather than 80 miles and vehicle efficiency increases by 15 percent, from 3.7 to 4.2 miles/kWh.

Brownstone et al. Brownstone et al. (1996) use the same 1993 household survey data as Kavalec and also use a multinomial-logit based model of household vehicle choice. They differ from him, however, in that they appear to use plausible assumptions about the attributes of EVs that will be available in 2003.[4] The main exceptions might be an overly optimistic assumption about the public recharging capability (10 percent of service stations are assumed to offer recharging) and an emphasis on 0-30 as opposed to 0-60 mph acceleration (EVs likely compare better with ICEVs on 0-30 mph acceleration times).

Brownstone et al. (p. 122) project that the full-function EV market share will amount to 5 percent when full-function EVs are priced at the same levels as comparable ICEVs. Market share remains at 3.6 percent even when full-function EVs carry a $10,000 premium.

---

[1]CALCARS (California Light Duty Vehicle Conventional and Alternative Fuel Response Simulator) is the forecasting methodology used by the California Energy Commission to project the number and types of personal cars and light-duty trucks owned, annual vehicle miles traveled, and fuel consumption in California.

[2]To account for battery replacement costs, EV operating costs are $0.07 per mile above those for a comparable ICEV. This amounts to $9,100 (undiscounted) over a 130,000-mile vehicle life, which is roughly equal to the cost of a replacement NiMH battery pack.

[3]Kavalec's assumed EV efficiency is 3.69 mi/kWh AC (wall-to-wheels), which is above the 2 to 3 range observed for most current vehicles. Top speed is assumed to be 94 mph, which is higher than that of most current EVs. Recharging time at service stations is assumed to be seven minutes, which even with fast charging systems is not currently feasible, and recharge time at home is three hours—shorter than the four to six hours required even with 220-volt lines. Finally, the ratio of public recharging facilities to gas stations is assumed to be 0.16, which is likely far in excess of the current public EV infrastructure in California. CARB (2000b, p. 61) estimates that there are currently 400 public charging stations in California, which offer about 700 chargers.

[4]They assume that range varies from 80 to 120 miles across the types of EVs available to consumers, top speed varies from 64 to 92 mph, operating costs are $0.07 per mile above those of gasoline vehicles, and it takes four hours to recharge an EV at home (Brownstone et al., 1996, p. 120).

Interpretation. The above studies suggest that the lifecycle costs of full-function EVs must be similar to those of comparable ICEVs in order to achieve 5 to 10 percent market shares. These full-function EVs have an 80 to 100 mile range and are similar in size, body style, styling, and amenities to the ICEVs used for comparison. There is evidence that some consumers are willing to pay a premium for EVs, but the number of these consumers is not large. This means that consumer surplus will accrue to some EV consumers at the market-clearing EV price, but this situation is not different from that in the ICEV market. Based on the results of these studies and the disappointing reception of EVs offered at price premiums between 1997 and 2000, we think it reasonable to conclude that the lifecycle cost of full-function EVs must not exceed those of comparable ICEVs if full-function EVs are to achieve meaningful market shares.

E.2 WILLINGNESS TO PAY FOR CITY ELECTRIC VEHICLES

We base our analysis of consumer willingness to pay for city EVs on the sales experience for small cars in the United States. As far as we know, no stated preference studies on the demand for city EVs are publicly available.

While no city EVs have ever been offered for sale in the United States, experience with sales of the smallest conventional cars may be instructive. The two smallest classes of cars sold domestically (minicompacts and subcompacts) have seen their market shares decline for two decades (Figure E.1).[5] Not only have the market shares declined, but the types of vehicles in each class have changed in ways that do not favor city EVs. Some minicompacts of the 1980s, such as the Renault Le Car and the Honda Civic CVCC, did functionally resemble city cars, but by model year (MY) 1990, all minicompacts were high-performance sports cars.

Table E.2 lists the smallest cars (by interior volume) currently available, excluding high-performance sports cars and luxury cars.[6] It shows that the domestically available vehicles most closely resembling city EVs sell for approximately $10,000 to $12,000. In total, these models have a market share of 0.5 percent, which corresponds to 7,500 vehicles when light-duty vehicle (LDV) sales are 1.5 million (roughly the current level in California). In our city EV production scenario (see Figure 3.3), sales (for California only) start below this level but soon exceed it, rising to over 10,000 per year by 2012. We think it unlikely that manufacturers will be able to sell this volume of city EVs if lifecycle costs exceed those of the small ICEVs on the road today.

[5]Minicompacts have interior volume less than 85 cubic feet; subcompacts are between 85 and 100 cubic feet.

[6]We do not list the prices for models with engines larger than 1.6 liters, which do not conform to the idea of a "very small car."

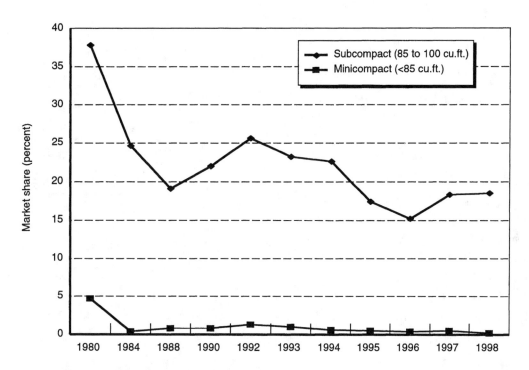

SOURCE: Oak Ridge National Laboratory,1998.

**Figure E.1—U.S. Market Share of Minicompact and Subcompact
Passenger Cars (excluding two-seaters)**

Table E.2

Passenger Cars with Smallest Interior Volume

Make/Model	Style	Passenger volume (cuft)	Luggage volume (cuft)	Interior volume (cuft)	Engine (liters)	Price[a]
Toyota Paseo	convertible	64	7	71	1.5	b
Toyota Celica	convertible	67	7	74	2.2	c
Mitsubishi Eclipse	hatchback	74	5	79	2.0	c
Toyota Paseo	2-door	74	8	82	1.5	b
Chevrolet Metro	hatchback	77	8	85	1.0/1.3	11,500
Suzuki Swift	hatchback	77	8	85	1.3	10,600
Honda Prelude	2-door	78	9	87	2.2	c
Toyota Celica	2-door	78	10	88	2.2	c
Acura Integra	hatchback	77	13	90	1.8	c
Chevrolet Metro	4-door	80	10	90	1.0/1.3	11,500
Volkswagen Cabrio	2-door	82	8	90	2.0	c
Hyundai Tiburon	2-door	80	13	93	2.0	c
Ford Escort	2-door	81	12	93	1.6	13,600
Nissan Sentra	2-door	84	10	94	1.6	13,200

SOURCE: U.S. Environmental Protection Agency, 2000.

[a]MY00 base price with standard transmission, air conditioning, stereo.

[b]Discontinued.

[c]Prices not listed because engine larger than 1.6 liters.

REFERENCES[1]

American Petroleum Institute, Facts About Oil, http://www.api.org/edu/factsoil.htm (viewed January 21, 2001).

Anderman, Menahem, Fritz Kalhammer, and Donald MacArthur, "Advanced Batteries for Electric Vehicles: An Assessment of Performance Cost and Availability," Battery Technology Advisory Panel of the California Air Resources Board, June 22, 2000 (often called the "BTAP report").

Arthur D. Little, Inc. (ADL), *Cost Analysis of Fuel Cell System for Transportation: Baseline System Cost Estimate*, prepared by Eric J. Carlson and Stephan Mariano, http://www.ott.doe.gov/pdfs/baseline_cost_model.pdf, March 2000.

Arthur D. Little Inc. (ADL), *Projected Automotive Fuel Cell Use in California*, prepared by Louis Browning, www.energy.ca.gov/reports/2002-02-06_P600-01-022F.PDF, October 2001.

Atkins, Lance, Tetsuyuki Taniguchi, Yasufumi Sakamoto, Dean Case, Rick Reinhard, and Hideki Shibayama, "Nissan Li-Ion Altra EV California Demonstration Status Report," Electric Vehicle Association of Asia Pacific, April 1999.

Automotive Intelligence News, "Ford of Canada Delivers First Th!nk City Electric Vehicle in Canada to Ballard Power Systems," September 21, 2000, www.autointell.com/news-2000-2/September-2000/September-26-00-p7.html (viewed October 31, 2001).

Ballard Power Systems, "The Ballard Fuel Cell," http://www.ewh.ieee.org/reg/7/millinnium/ballard/ballard_info.html, 2002.

Bartis, James T., RAND, personal communication, November 2001.

Bevilacqua-Knight, Inc., *Bringing Fuel Cell Vehicles to Market: Scenarios and Challenges with Fuel Alternatives*, Hayward, CA, October 2001.

Brooks, Alec, and Tom Gage, "Integration of Electric Drive Vehicles with the Electric Power Grid—a New Value Stream," *Proceedings of the 18th International Electric Vehicle Symposium*, Berlin, October 2001.

Brown, Mark B., Weert Canzler, Frank Fischer, and Andreas Knie, "Technological Innovation Through Environmental Policy: California's Zero-Emission Vehicle Regulation," *Public Productivity and Management Review*, Vol. 19, No. 1, pp. 77–93, 1995.

Brown, James, "First impressions of the Th!nk City," August 18, 2001, evworld.com (viewed October 31, 2001).

Brownstone, David, David S. Bunch, Thomas F. Golob, and Weiping Ren, "A Transactions Choice Model for Forecasting Demand for Alternative-Fuel Vehicles," *Research in Transportation Economics*, Vol. 4, pp. 87-129, JAI Press Inc, 1996.

---

[1]Some materials, especially those appearing only on Web sites, are undated. To reference these materials in the text, we used the viewing dates.

Bruck, Rolf, Peter Hirth, and Wolfgang Maus, "The Necessity of Optimizing the Interactions of Advanced Post-Treatment Components in Order to Obtain Compliance with SULEV-Legislation," SAE paper 1999-01-0770, 1999.

Brungard, Terry, "Hypermini Finds Utility Niche," *EV World*, www.evworld.com/databases/storybuilder.cfm?storyid=221 (viewed November 5, 2001).

Bulk Molding Compounds, Inc., "BMC, Inc. Signs Exclusive License Agreement with Los Alamos National Laboratory," *BMC, Inc. Briefs*, Vol. 13, No. 2, West Chicago, IL, October, 2001.

Burke, A.F., K.S. Kurani, and E.J. Kenney, *Study of the Secondary Benefits of the ZEV Mandate*, Institute of Transportation Studies, UC Davis, August 2000.

California Academy of Sciences, http://www.calacademy.org/events/earthday2001/fossil_fuel.html (viewed November 2001).

California Air Resources Board (CARB), *The California State Implementation Plan for Ozone, Volume II: The Air Resources Board's Mobile Source and Consumer Products Elements*, November 15, 1994.

California Air Resources Board, *Staff Report: Public Hearing to Consider Amendments to the 1999 Small Off-Road Engine Regulations*, February 6, 1998a.

California Air Resources Board, *Staff Report: Initial Statement of Reasons; Proposed Amendments to Heavy-Duty Vehicle Regulations: 2004 Emission Standards; Averaging, Banking and Trading; Optional Reduced Emission Standards; Certification Test Fuel; Labeling; Maintenance Requirements and Warranties*, March 6, 1998b.

California Air Resources Board, *Staff Report: Public Hearing to Consider Adoption of Emission Standards and Test Procedures for New 2001 and Later Off-Road Large Spark-Ignition Engines*, September 4, 1998c.

California Air Resources Board, *Staff Report: Initial Statement of Reasons; Public Hearing to Consider Amendments to Off-Road Compression-Ignition Engine Regulations: 2000 and Later Emission Standards, Compliance Requirements and Test Procedures*, December 10, 1999a.

California Air Resources Board, *EMFAC2000, Version 1.99f (Beta)*, downloaded from http://www.arb.ca.gov/msei/msei.html, November 1999b.

California Air Resources Board, *Initial Statement of Reasons for the Proposed Amendments to the Regulation for Reducing Volatile Organic Compound Emissions from Aerosol Coating Products and Proposed Tables of Maximum Incremental Reactivity (MIR) Values, and Proposed Amendments to Method 310, Determination of Volatile Organic Compounds in Consumer Products*, Chapter XI, May 5, 2000a.

California Air Resources Board, *Staff Report: 2000 Zero Emission Vehicle Program Biennial Review*, August 7, 2000b.

California Air Resources Board, *Staff Report: Initial Statement of Reasons; Proposed Amendments to the California Zero-Emission Vehicle Program Regulations*, December 8, 2000c.

California Air Resources Board, letter to Felicia Marcus, Administrator, Region IX, U.S. EPA, August 17, 2000d.

California Air Resources Board, *ARB Fact Sheet: Zero Emission Vehicle Program Changes*, February 23, 2001a.

California Air Resources Board, *Staff Report: Initial Statement of Reasons; Public Hearing to Consider Amendments Adopting More Stringent Emission Standards for 2007 and Subsequent Model Year New Heavy-Duty Diesel Engines*, September 7, 2001b.

California Air Resources Board, *Status of State and Federal Emission Reductions Commitments for 2010 in the South Coast's 1999 SIP*, memo supplied by Andrew Panson, Planning and Technical Support, October 25, 2001c.

California Air Resources Board, *ARB Staff Review of Report Entitled "Impacts of Alternative ZEV Sales Mandates on California Motor Vehicle Emissions: A Comprehensive Study,"* October 31, 2001d.

California Air Resources Board, *Proposed Regulation Order Modified Text, Amendment to California Zero Emission Vehicle Regulation—Section 1962, Title 13, California Code of Regulation—and Related Provisions*, October 31, 2001e.

California Air Resources Board, *Re: Denial of General Motors Corporation's Request for Administrative Action Under Government Code Section 11340.6 Regarding the California Zero Emission Vehicle Regulation*, petition decision, March 9, 2001f.

California Air Resources Board, *California Exhaust Emission Standards and Test Procedures for 2001 and Subsequent Model Passenger Cars, Light-Duty Trucks, and Medium-Duty Vehicles*, September 28, 2001g.

California Air Resources Board, *Amendments to the California Zero Emission Vehicle Program Regulations, Final Statement of Reasons*, December 2001h.

California Air Resources Board, *Final Regulation Order, Amendments to the California Zero Emission Vehicle Regulation—Section 1962, Title 13, California Code of Regulations—and Related Provisions*, April 12, 2002.

California State University at Fullerton, Institute for Economic and Environmental Studies, memorandum to Chuck Schulock at the California Air Resources Board, October 15, 2001.

California ZEV Alliance, "The Sky is Falling," www.zevnow.org/sky.pdf (viewed December 9, 2001).

Chalk, Steven, "Overview of DOE Transportation Fuel Cell R&D," presentation, U.S. Department of Energy, February 14, 2002.

Connelly, Mary, "New Global Ford Group Th!nks Green," *Automotive News*, January 17, 2000, www.autonews.com (viewed March 22, 2000).

Cuenca, R., L. Gaines, and A. Vyas, *Evaluation of Electric Vehicle Production and Operating Costs*, Center for Transportation Research, Argonne National Laboratories, ANL/ESD-41, 1999.

Daniel, Mac, "Drive to Cleaner Driving Shifts into a Lower Gear," *Boston Globe*, January 8, 2002, www.boston.com/dailyglobe (viewed January 10, 2002).

Dayton, James, "PEM Bipolar Plate Fabrication," presented at NCMS Fall Workshop, http://techcon.ncms.org/99fall/presentations/FuelCells_Dayton.pdf, September 27, 1999.

Delucchi, Mark A., Andrew F. Burke, Timothy E. Lipman, and Marshall Miller, *Electric and Gasoline Vehicle Lifecycle Cost and Energy-Use Model: Report for the California Air Resources Board*, Institute of Transportation Studies, UC Davis, UCD-ITS-RR-99-4, April 2000.

Directed Technologies, Inc. (DTI), *Detailed Manufacturing Cost Estimates for Polymer Electrolyte Membrane (PEM) Fuel Cells for Light Duty Vehicles*, technical report prepared for Ford Motor Company, Arlington, VA, August 1998.

Dixon, Lloyd, and Steven Garber, *California's Ozone-Reduction Plan for Light-Duty Vehicles: Direct Costs, Direct Emission Effects, and Market Responses*, RAND, Santa Monica, CA, MR-695-ICJ, 1996.

Dixon, Lloyd, and Steven Garber, *Fighting Air Pollution in Southern California by Scrapping Old Vehicles*, RAND, Santa Monica, CA, MR-1256-PPIC/ICJ, 2001.

Duleep, K.G., "Briefing on Technology and Cost of Toyota Prius," unpublished report prepared by Energy and Environmental Analysis for U.S. DOE Office of Transportation Technologies, 1998.

Easterbrook, Greg, "Cleaning Up," *Newsweek*, pp. 29–30, July 24, 1989.

Electric Vehicle Association of America (EVAA), www.evaa.org/evaa/pages/ele_product_info.htm (viewed November 5, 2001).

Energy Information Administration, monthly report, http://www.eia.doe.gov/emeu/ipsr/contents.html (viewed November 2001).

EV World, "*Whither Tomorrow's Cars? Synopsis of a Blue Ribbon Panel Discussion at 1999 Environmental Vehicle Conference*," Ypsilanti, MI, June 14-16," http://evworld.com/archives/conferences/env99/env99_blueribbon. html, 1999.

EV World, evworld.com/archives/currents003.html (viewed March 3, 2000).

Ford Motor Company, "Ford to Introduce Th!nk City EV in First North American Demonstration Program," press release, December 1, 1999, www.thinkmobility.com (viewed December 1, 2001).

Fuel Cell Catalyst, Vol. 2, No. 1, Fall 2001.

Gas Turbine World, *Gas Turbine Handbook, 2000-2001.*

Green Car Institute, *Future EV Pricing*, www.greencars.com, 2001.

Greene, David L., James R. Kahn, and Robert C. Gibson, "Fuel Economy Rebound Effect for U.S. Household Vehicles," *The Energy Journal*, Vol. 20, No. 3, pp. 1-31, 1999.

Gruenspecht, Howard K., "Differentiated Regulation: The Case of Auto Emission Standards," *American Economic Review*, Vol. 72, pp. 328-331, 1982a.

Gruenspecht, Howard K., *Differentiated Regulation: A Theory with Applications to Automobile Emissions Controls*, Ph.D. dissertation, Yale University, 1982b.

Gruenspecht, Howard, "Zero Emission Vehicles: A Dirty Little Secret," *Resources*, Issue 142, 7, 2001.

Harrison, David, Jr., Daniel Radov, Philip Heirings, James Lyons, Thomas Austin, Bernard Reddy, and Michael Lovenheim, *Impacts of Alternative ZEV Sales Mandates on California Motor Vehicle Emissions: A Comprehensive Study*, National Economic Research Associates and Sierra Research, January 2001.

Hart, Dan, "Toyota to Start Selling Electric RAV4 to Consumers in February," *Bloomberg News*, December 13, 2001.

Hill, Daniel H., "Derived Demand Estimation with Survey Experiments: Commercial Electric Vehicles," *Review of Economics and Statistics*, pp. 277-285, May 1987.

Honda Motor Company, *Honda Introduces New Fuel Cell-Powered Vehicle, FCX-V4*, Tokyo, September 4, 2001.

Jones, Terril Yue, and John O'Dell, "Auto Makers to Form Fuel-Cell Partnership," *Los Angeles Times*, C3, January 9, 2002.

Kalhammer, Fritz, Paul Prokopius, Vernon Roan, and Gerald Voecks, *Status and Prospects of Fuel Cells as Automobile Engines*, available at http//www.arb.ca.gov/msprog/zevprog/fuelcell/kalhamm7, August 27, 1999 (often called the "FCTAP study").

Kassakian, John G., "Automotive Electrical Systems: The Power Electronics Market of the Future," *Proceedings of the IEEE Applied Power Electronics Conference*, New Orleans, pp. 3-9, February 2000.

Kavalec, Chris, *CALCARS: The California Conventional and Alternative Fuel Response Simulator*, California Energy Commission, P500-96-003, 1996.

Korthof, "Level 2+: Economical Fast Charging for EVs," presented at the 17th Electric Vehicle Conference (EVS17), Montreal, Canada, October, 2000.

Lipman, Timothy, *A Dynamic, Fuzzy Set-Based Framework for Assessing the Social Costs of Emerging Electric-Drive Vehicle Technologies*, Ph.D. dissertation, UC Davis, 1999a.

Lipman, Timothy, *The Cost of Manufacturing Electric Vehicle Drivetrains*, Institute of Transportation Studies, UC Davis, UCD-ITS-RR-99-7, 1999b.

Lipman, Timothy, Mark Delucchi, and David Friedman, *A Vision of Zero-Emission Vehicles: Scenario Cost Analysis from 2003 to 2030*, analysis and report prepared for the Steven and Michelle Kirsch Foundation and the Union of Concerned Scientists, final report, October 23, 2000.

Los Alamos National Laboratory (LANL) http://www.uscar.org/pngv/technical/labs.htm (viewed January 2002).

Mace, David, "Electric Car Requirement Short-Circuited," *Rutland Herald,* January 2, 2002 (viewed at rutlandherald.nybor.com/News/State/Story/39497.html).

Matsushita Electric Industrial Co, Ltd., "Matsushita Electric (Panasonic) To Launch Lead-Acid Battery for Advanced Electric Vehicles (EV)," Matsushita Battery Industrial, Co., Ltd, Tokyo, July 19, 2000.

Moore, Bill, "Ford Buys Th!nk EV Manufacturer," *EV World*, evworld.com/reports/ford-th!nk.html, viewed March 27, 2000a.

Moore, Bill, "Congress Focuses on PNGV Hybrid Electrics," *EV World*, evworld.com/conferences/futurecar2000/bmoore3.html (viewed April 10, 2000b).

National Economic Research Associates, "Impacts of the Zero Emission Mandate on the California Economy," prepared for General Motors, submitted with "Response to Notice of Public Hearing to Consider Amendments to the California Zero Emission Vehicle Regulations Published on December 8, 2000," Vol. II, Tab 2, 2000.

National Research Council, *Effectiveness and Impact of Corporate Average Fuel Economy (CAFE) Standards*, Committee on the Effectiveness and Impact of Corporate Average Fuel Economy (CAFE) Standards, Board on Energy and Environmental Systems, Transportation Research Board, National Research Council, National Academy of Sciences, 2002.

New Car Test Drive, "2003 Think City," www.newcartestdrive.com/sneakpreviews/03thinkcity_sp.cfm (viewed November 7, 2001).

Nissan, "Start of Joint Field Demonstration Project Using Nissan Hypermini," press release, January 13, 2000, www.nissan-global.com/GCC/Japan/NEWS/20000113_0e.html (viewed November 5, 2001).

Nissan, "Nissan's Hyperminis in UC Davis Research: Project Focused on Marketability of Electric 'City Cars'," press release, November 7, 2001, www.nissanews.com (viewed November 7, 2001).

Oak Ridge National Laboratory, "Light Vehicles and Characteristics," in *Transportation Energy Data Book*, Edition 18, ORNL-6941, 1998.

Oei, D., J.A. Adams, A.A. Kinnelly, G.H. Purnell, R.I. Sims, M.S. Sulek, and D.A. Wernett from Ford Motor Company, and B. James, F. Lomax, G. Baum, S. Thomas, and I. Kuhn for Directed Technologies, Inc., *Direct Hydrogen Fueled Proton Exchange Membrane Fuel Cell System for Transportation Applications: Final Technical Report,* Ford Motor

Company, Dearborn, MI, U.S. Department of Energy report DOE/CE/50389-595, December 2000.

Ogden, Joan M., *Review of Small Stationary Reformers for Hydrogen Production,* Center for Energy and Environmental Studies, Princeton University, NJ, 2001.

Parry, Ian W.H., and Kenneth A. Small, *Does Britain or the United States Have the Right Gasoline Tax?* Resources for the Future, discussion paper 02-12, March 2002.

Percival, Robert V., Alan S. Miller, Christopher H. Schroeder, and James P. Leape, *Environmental Regulation: Law, Science, and Policy*, Little, Brown, and Co., Boston, MA, 1992.

Perez-Peña, Richard, "Zero-Emission Quota on Cars Is Postponed for Two Years," *The New York Times,* B5, January 4, 2002.

Phelan, Mark, "Honda Plans to Lead in U.S. with Hydrogen Fuel Cell Car," *Detroit Free Press,* March 4, 2002.

PRNewswire, "Advanced Automotive Batteries for Hybrid Electric Vehicle Battery Market Charged for Rapid Growth Reports New Study," September 10, 2001.

Reed, Leslie Harrison, Jr., "California Low-Emission Vehicle Program: Forcing Technology and Dealing Effectively with the Uncertainties," *Boston Coll. Environmental Affairs Law Review*, Vol. 24, No. 4, pp. 695-793, 1997.

Reinhold, Robert, "With Help from Local Governments, U.S. Industry Tries to Remake Itself," *The New York Times*, B8, March 2, 1992.

Reuyl, John S., and Pierre J. Schuurmans, *Policy Implications of Hybrid-Electric Vehicles,* NEVCOR, Inc., Stanford, CA, April 22, 1996.

Schreffler, Roger, "Toyota Shines at Tokyo Show with Gasoline Fuel Cell SUV," *Hydrogen & Fuel Cell Letter,* November 2001, http://www.hfcletter.com/letter/November01/feature.html (viewed January 19, 2002).

Schwartz, Joel, *Smog Check II Evaluation, Part IV: Smog Check Costs and Cost Effectiveness*, California Inspection and Maintenance Review Committee, Sacramento, 2000.

Singapore Straits Times, "Electric Cars Expected on the Roads by June," March 28, 2001, straitstimes.asia1.com.sg/singapore/story/0,1870,32777,00.html, archived at www.geocities.com/sjeaanews/2001/5/page1.html (viewed November 5, 2001).

South Coast Air Quality Management District (SCAQMD), *1997 Air Quality Management Plan*, Diamond Bar, CA, November 1996.

South Coast Air Quality Management District, *1999 Amendments to the 1997 Ozone State Implementation Plan for the South Coast Air Basin*, final, Diamond Bar, CA, 1999.

South Coast Air Quality Management District, *Historic Ozone Air Quality Trends*, http:///www.aqmd.gov/smog/o3trend.html (viewed January 7, 2001).

Southern California Edison, *Schedule TOU-EV-1, Domestic Time-of-Use Electric Vehicle Charging,* filed December 22, 2000, Rosemead, CA.

Stewart, Kenneth C., *Presentation to the Air Resources Board,* General Motors, Detroit MI, May 31, 2000.

Sullivan, Bob, "New Battery a Quiet Auto Revolution," http://www.msnbc.com/news/674071.asp (viewed December 17, 2001).

Thomas, C., I. Kuhn, B. James, F. Lomax, and G. Baum, "Affordable Hydrogen Pathways for Fuel Cell Vehicles," *International Journal of Hydrogen Energy*, Vol. 23, No. 6, pp. 507-516, 1999.

Thomas C.E., Brian D. James, Franklin D. Lomax, Jr., and Ira F. Kuhn, Jr., "Fuel Options for the Fuel Cell Vehicle: Hydrogen Methanol or Gasoline?" presented at the Fuel Cell Reformer Conference, South Coast Air Quality Management District, Diamond Bar, CA, November, 20, 1998.

Tietenberg, Tom, *Environmental and Natural Resource Economics,* third edition, Harper Collins Publishers, Inc., 1992.

Tribdino, Raymond B., "Nissan Hypermini EV," *EV World,* evworld.com/archives/testdrives/hypermini.html (viewed October 31, 2001).

Turrentine, Thomas, and Kenneth Kurani, *The Household Market for Electric Vehicles: Testing the Hybrid Household Hypothesis—A Reflexively Designed Survey of New-Car-Buying, Multi-Vehicles California Households,* Institute for Transportation Studies, UC Davis, UCD-ITS-RR-95-5, May 12, 1995.

UC Davis, press release, www.pubcomm.ucdavis.edu/newsreleases/11.01/news_hypermini_www.html (viewed November 5, 2001).

UC Irvine, press release, www.apep.uci.edu/ecom/purpose.html (viewed November 5, 2001).

Unnasch, Stefan, "Fueling Station Infrastructure Cost Analysis," in Bevilacqua-Knight, Inc., *Bringing Fuel Cell Vehicles to Market: Scenarios and Challenges with Fuel Alternatives,* Hayward, CA, October 2001.

Unnasch, Stefan, Louis Browning, and Michelle Montano, *Evaluation of Fuel-Cycle Emissions on a Reactivity Basis*, ARB contract A166-134, Acurex Environmental, Project 8522, 1996.

U.S. Department of Energy, www.ott.doe.gov/biofuels/energy.html, 2001.

U.S. Environmental Protection Agency, *Model Year 2000 Fuel Economy Guide*, 2000.

U.S. Supreme Court, *Whitman, Administrator of Environmental Protection Agency, et al. v. American Trucking Associations, Inc., et al.,* 99-1257, February 26, 2002.

USA Today, "Ford Unveils Electric Vehicles," January 10, 2000, www.usatoday.com/life/cyber/technology/review/crg795.htm.

Vyas, A., R. Cuenca, and L. Gaines, *An Assessment of Electric Vehicle Life Cycle Costs to Consumers*, SAE technical paper 982182, 1998.

Wallace, John, "Ford Re-Th!nks Its EV Vision," interview by *EV World*, audio at evworld.com/interviews/wallace.html (heard March 27, 2000).

Walsh, Michael, "Norway Opens Electric Car Plant," *Car Lines*, Vol. 99, No. 6, December 1999.

Ward's Communications, *Motor Vehicle Facts and Figures 1999,* Southfield, MI, 1999.

Wolff, G., D. Rigby, D. Gauthier, and M. Cenzatti, "The Potential Impacts of an Electric Vehicle Manufacturing Complex on the Los Angeles Economy," *Environment and Planning A*, Vol. 27, No. 6, pp. 877-905, 1995.